What does this painting have to do with math?

Color and music fascinated Wassily Kandinsky, an abstract painter and trained musician in piano and cello. Some of his paintings appear to be "composed" in a way that helps us see the art as a musical composition. In math, we compose and decompose numbers to help us become more familiar with the number system. When you look at a number, can you see the parts that make up the total?

On the cover

Thirteen Rectangles, 1930
Wassily Kandinsky, Russian, 1866–1944
Oil on cardboard
Musée des Beaux-Arts, Nantes, France

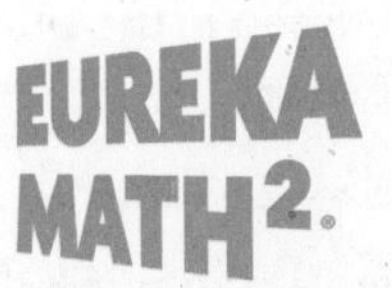

Great Minds® is the creator of *Eureka Math®*, *Wit & Wisdom®*, *Alexandria Plan™*, and *PhD Science®*.

Published by Great Minds PBC.
greatminds.org

Printed in the USA
B-Print

3 4 5 6 7 8 9 10 11 12 QDG 27 26 25 24 23

ISBN 978-1-64497-118-5

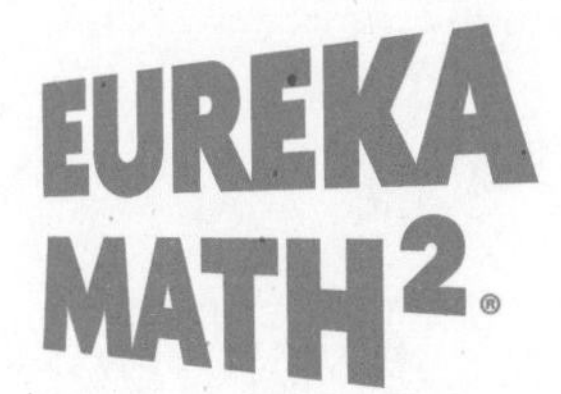

A Story of Units®

Fractions Are Numbers ▸ 5

LEARN

1 Place Value Concepts for Multiplication and Division with Whole Numbers

2 Addition and Subtraction with Fractions

3 Multiplication and Division with Fractions

4 Place Value Concepts for Decimal Operations

5 Addition and Multiplication with Area and Volume

Module 6 Foundations to Geometry in the Coordinate Plane

Contents

Foundations to Geometry in the Coordinate Plane

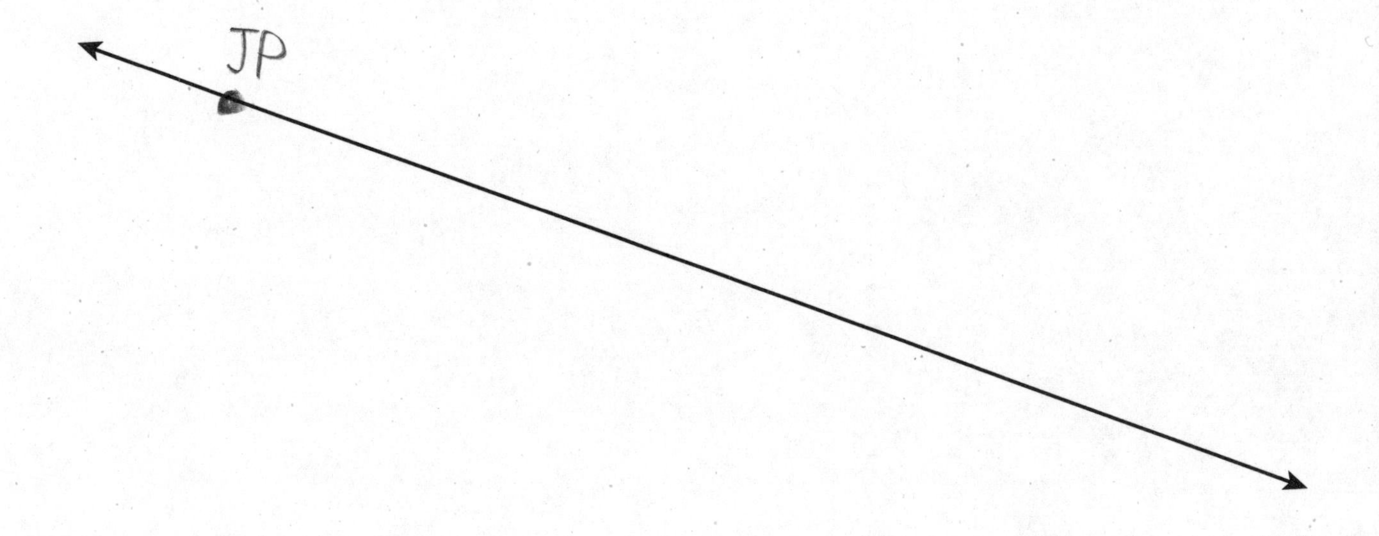
JP

1

Name _______________ Date _______

1. Use the line to create a coordinate system.

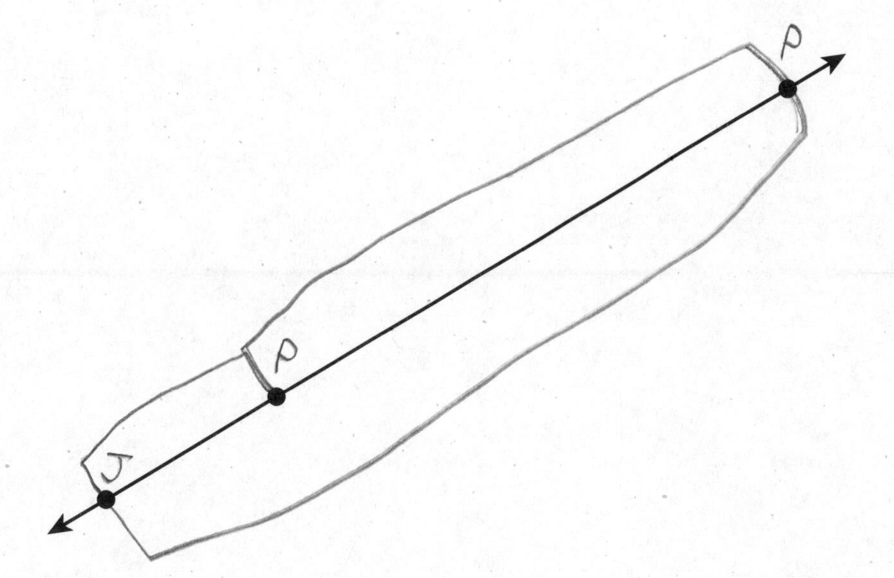

1

Name ______________________ Date ____________

Use the number line to complete problems 1–3.

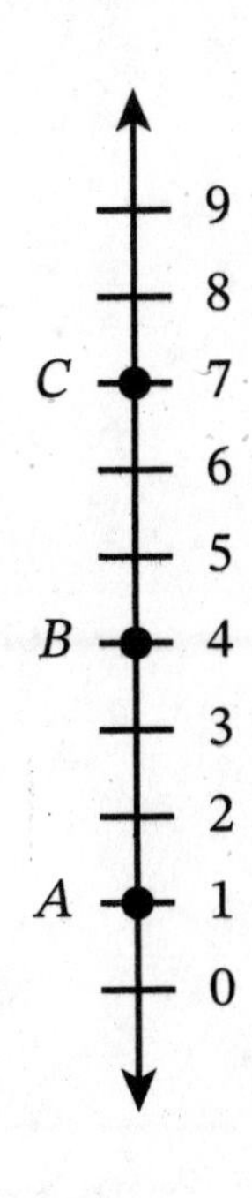

1. The coordinate of point A is __1__.

2. Point __B__ has a coordinate of 4.

3. The distance from point A to point C is __6__ units.

For problems 4 and 5, plot the point on the number line.

4. Plot point A so its distance from 0 is 2 units.

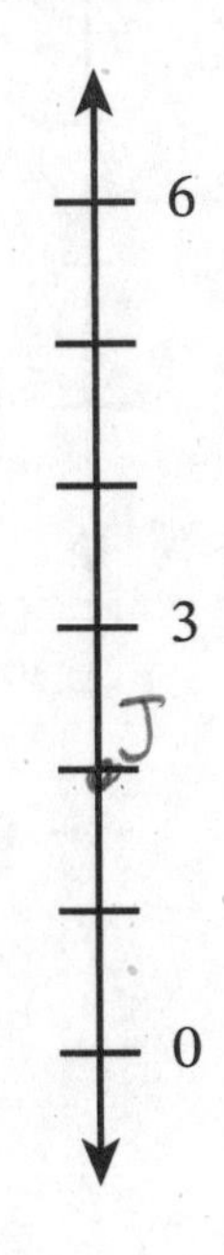

5. Plot point R so its distance from 0 is $\frac{5}{2}$ units.

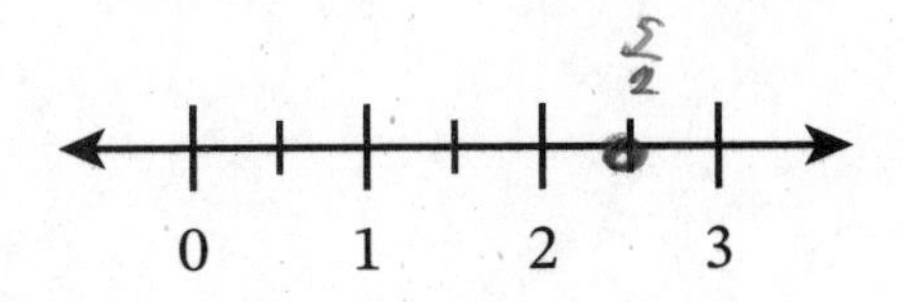

$2\frac{1}{2}$

6. Use the number line to complete parts (a)–(c).

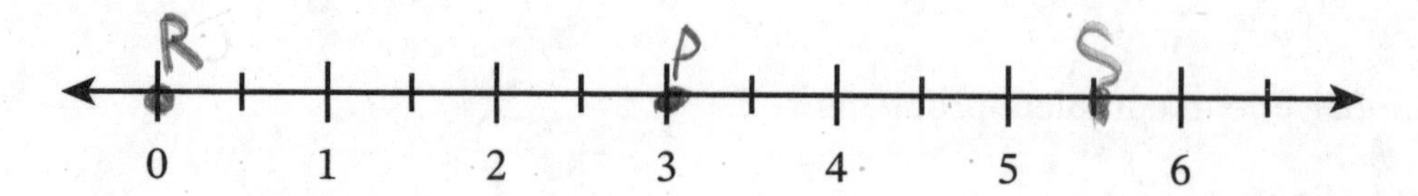

a. Plot and label point P at 3.

b. Plot and label point R at 0.

c. Plot and label point S so that it is $\frac{5}{2}$ units farther from 0 than point P. What is the coordinate of point S?

---

7. Plot and label point L so that its distance from 0 is 125 units.

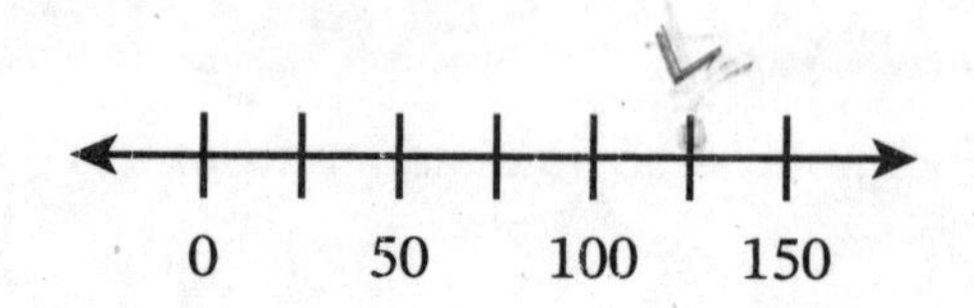

8. Plot point T so that it is $\frac{2}{3}$ units farther from 0 than point S.

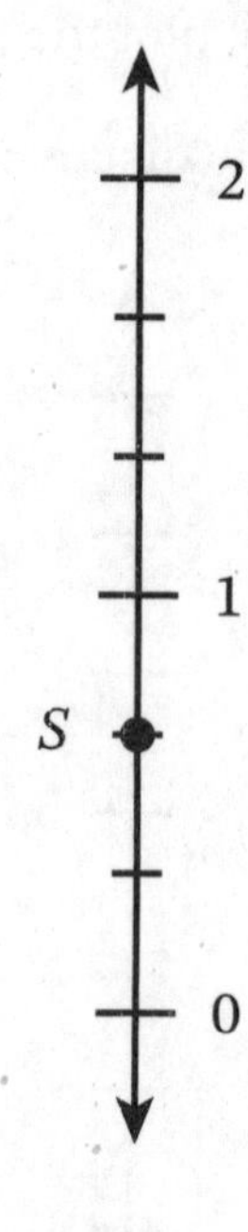

9. Construct a coordinate system on the line. Choose an interval length that allows each of the points described to be plotted. Plot and label the points.

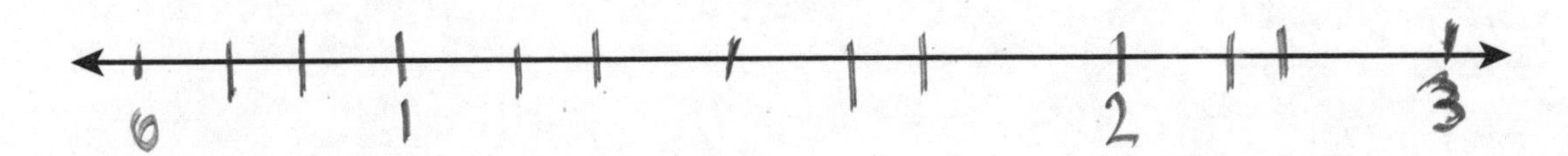

a. Point A is located 3 units from 0.

b. Point B is located at $\frac{2}{3}$.

c. Point C is located $1\frac{1}{3}$ units farther from 0 than point B.

d. Point D is located $\frac{2}{3}$ units closer to 0 than point A.

---

10. Blake asks Sasha to plot a point, P, that is 3 units from point M. Sasha says the coordinate of point P could be 1 or 7. How can Blake make the directions clear so that Sasha knows exactly where to plot point P? Explain.

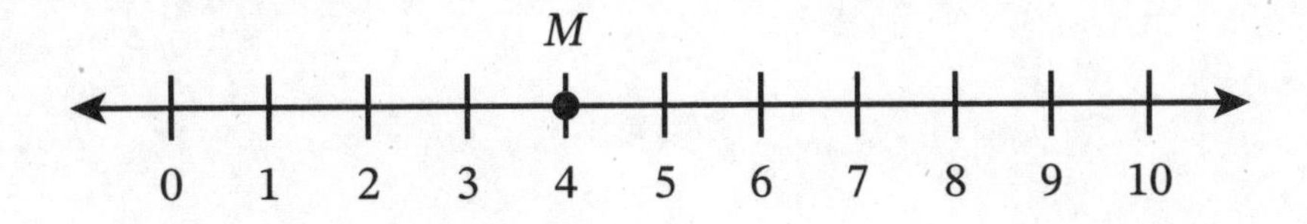

1

Name ______________________ Date ____________

Construct a coordinate system on the line. Choose an interval length that allows each of the points described to be plotted. Plot and label the points.

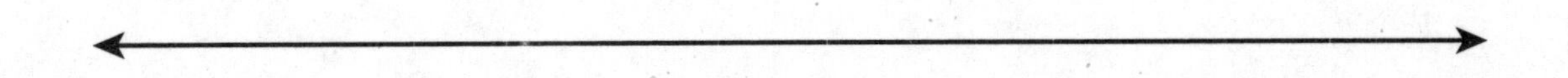

a. Point A is located 1 unit from 0.

b. Point B is located at $2\frac{1}{4}$.

c. Point C is located $\frac{1}{2}$ unit farther from 0 than point A.

d. Point D is located $\frac{1}{2}$ unit closer to 0 than point B.

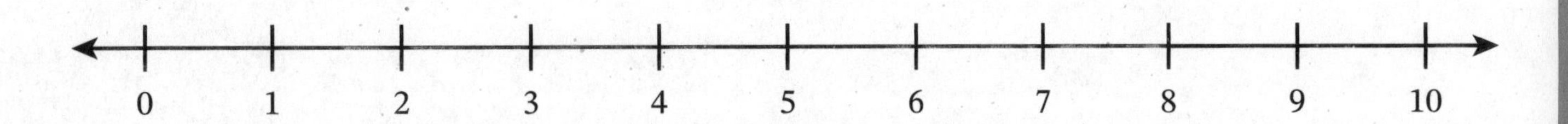
0
1
2
3
4
5
6
7
8
9
10

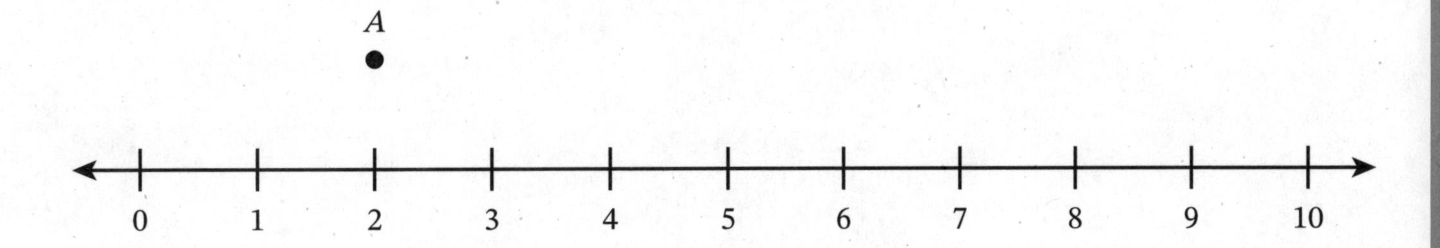
A
0
1
2
3
4
5
6
7
8
9
10

2

Name ______________________ Date __________

1. Use the coordinate plane to complete the problem.

   a. Label the axes x and y.

   b. Label the origin as 0.

   c. From the origin, label every grid line on both axes with a whole number from 1 to 10.

   d. Plot and label point P with x-coordinate 4 and y-coordinate 3.

   e. Plot and label point Q at (3, 4).

   f. Plot and label point R located 9 units to the right of the y-axis and 7 units above the x-axis.

   g. Plot and label point S at (7, 9).

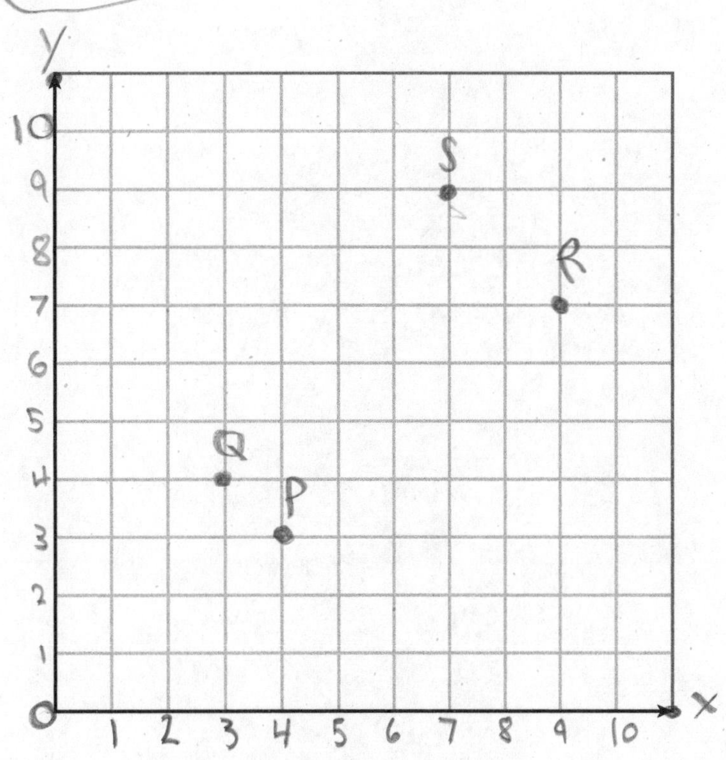

2. Complete the table for the points plotted in problem 1.

Point	x-Coordinate	y-Coordinate	Ordered Pair (x, y)
P	4	3	(4,3)
Q	3	4	(3,4)
R	9	7	(9,7)
S	7	9	(7,9)

2

Name ______________________ Date __________

Use the graph to complete problems 1–3.

1. Label the origin and the x- and y-axes.

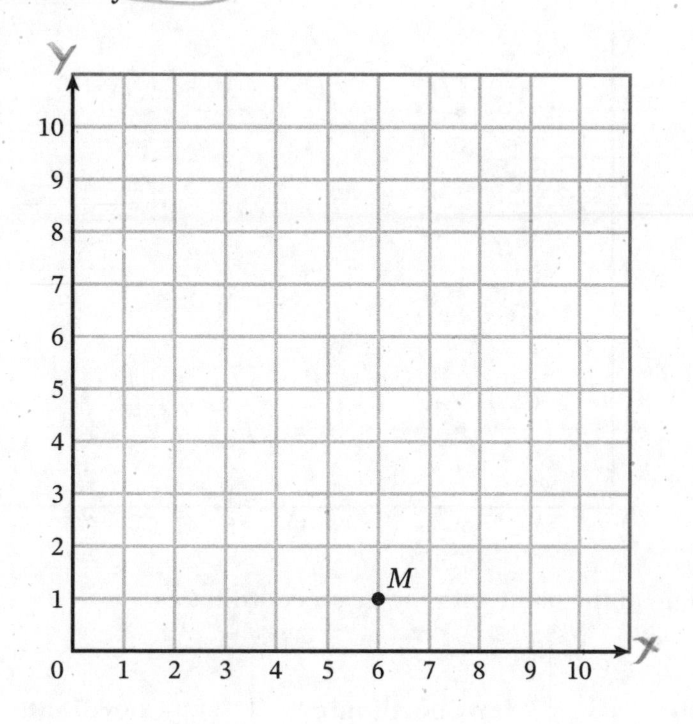

2. Consider point M.

   a. The x-coordinate of point M is 6.

   b. The y-coordinate of point M is 1.

   c. The ordered pair that identifies the location of point M is (6, 1).

3. What is the ordered pair that identifies the location of the origin?

(0,0)

4. Use the graph to complete parts (a)–(d).

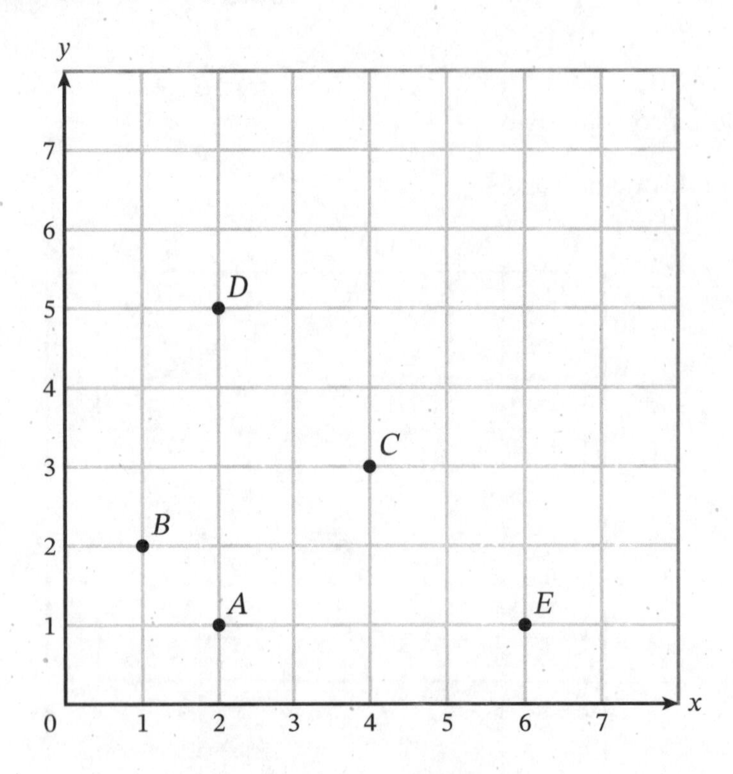

a. Write the name of the point with the given coordinates.

Point	x-Coordinate	y-Coordinate	Ordered Pair (x, y)
B	1	2	(1, 2)
A	2	1	(2, 1)
C	4	3	(4, 3)
E	6	1	(6, 1)

b. Point D is 5 units to the right of the y-axis.

c. Point D is 2 units above the x-axis.

d. Write the ordered pair that represents point D.

(5, 2)

5. Use the graph to complete parts (a)–(c).

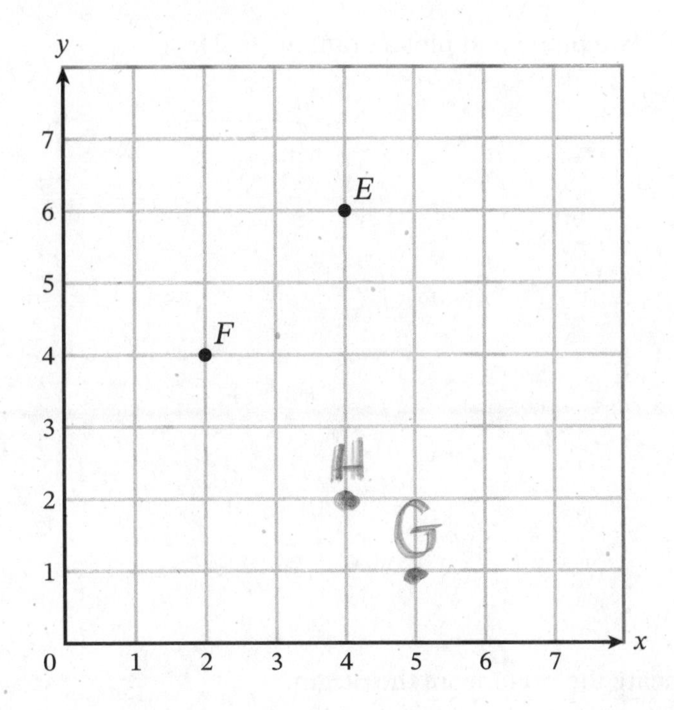

a. Complete the table for points E and F.

Point	x-Coordinate	y-Coordinate	Ordered Pair (x, y)
E	4	6	(4, 6)
F	2	4	(2, 4)

b. Plot a point at (5, 1). Label the point G.

c. Plot a point at (4, 2). Label the point H.

6. Consider the ordered pair (6, 2).

   a. Construct a coordinate plane and plot a point at (6, 2).

   b. Explain how to locate the point from the origin.

---

7. Yuna makes a mistake and says the point plotted at (6, 2) is located 2 units to the right of the y-axis and 6 units above the x-axis. Write a statement to correct Yuna's mistake.

E

F

B

D

C

A

Coordinate Plane A

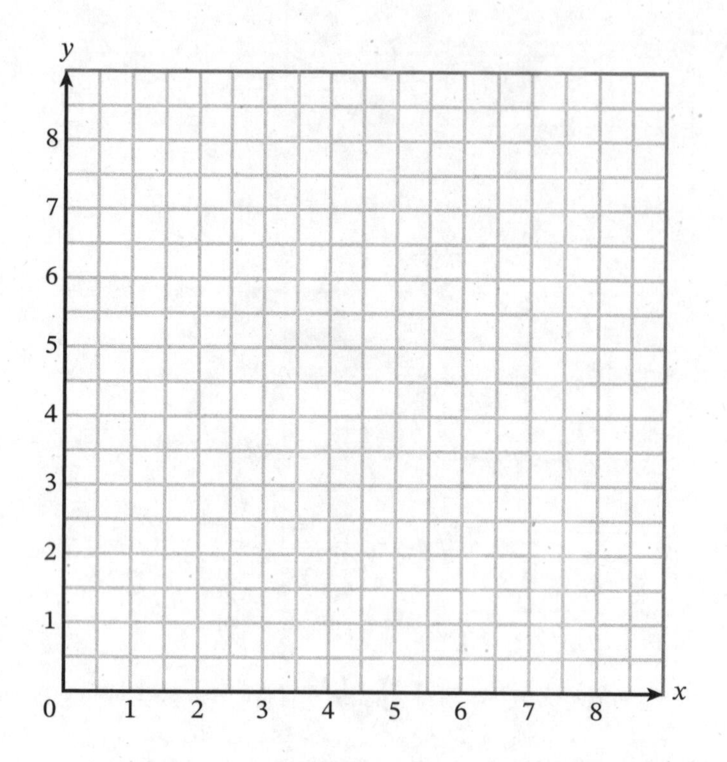

Coordinate Plane B

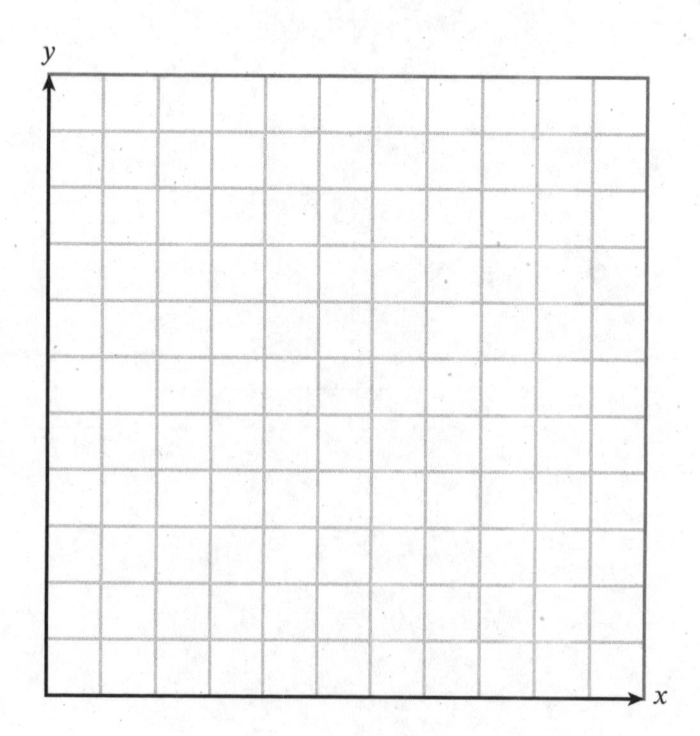

Name ____________________ Date __________

1. Write the coordinates and ordered pairs for points P, Q, R, S, and T in the table.

Point	x-Coordinate	y-Coordinate	Ordered Pair
P	0	$1\frac{3}{4}$	$(0, 1\frac{3}{4})$
Q	$1\frac{1}{4}$	$2\frac{3}{4}$	$(1\frac{1}{4}, 2\frac{3}{4})$
R	$1\frac{3}{4}$	0	$(1\frac{3}{4}, 0)$
S	$2\frac{1}{4}$	$\frac{3}{4}$	$(2\frac{1}{4}, \frac{3}{4})$
T	0	0	$(0, 0)$

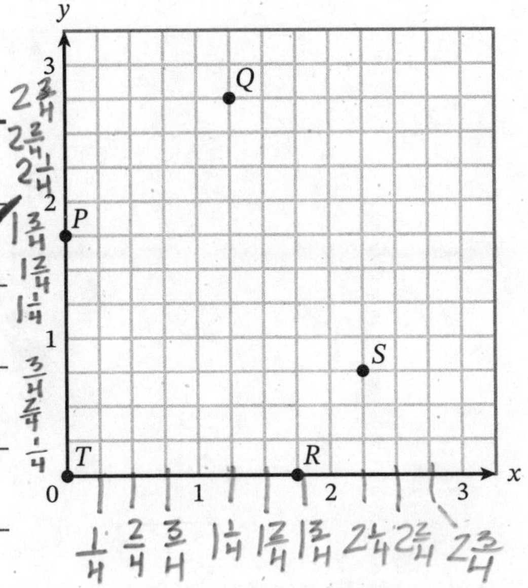

3

Name ______________________ Date __________

1. Use the graph to complete parts (a)–(e).

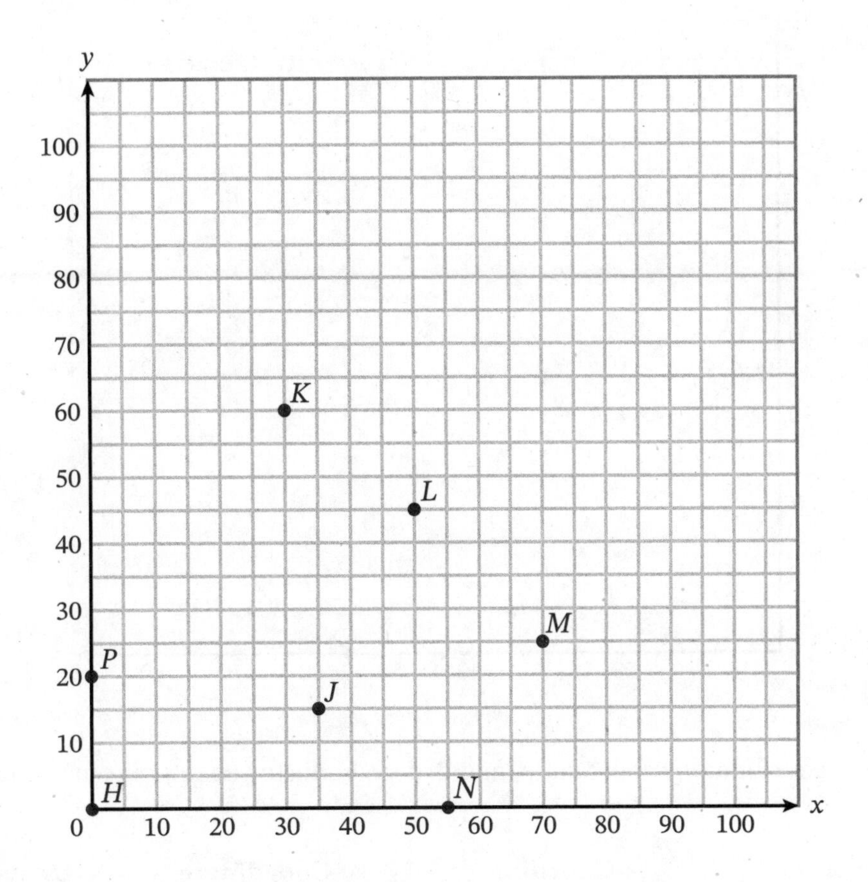

a. The ordered pair (35, 15) describes the location of point ______.

b. The ordered pair that describes the location of point K is (______, ______).

c. Point ______ is located at the origin.

d. Which two points have an x-coordinate of 0?

e. Which two points are located on the x-axis?

2. Use the graph to complete parts (a) and (b).

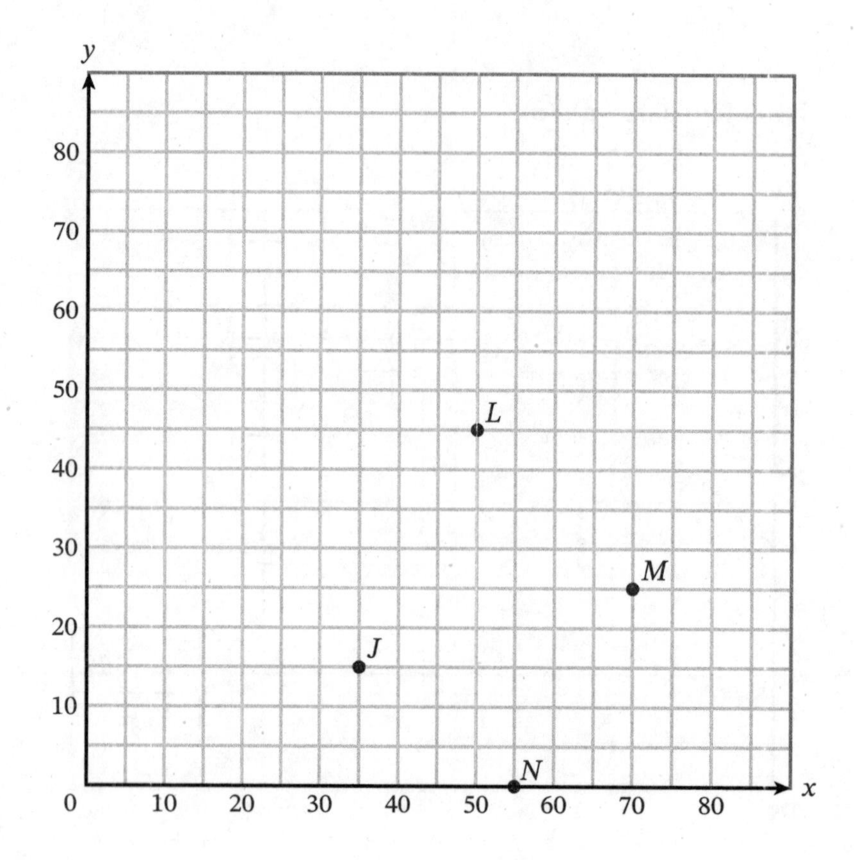

a. Write the x-coordinate, y-coordinate, and ordered pair for each point in the table.

Point	x-Coordinate	y-Coordinate	Ordered Pair
L			
M			
N			

b. Lacy says the ordered pair for point J is (31, 11). Is Lacy correct? Explain.

3. Use the coordinate plane to complete parts (a)–(e).

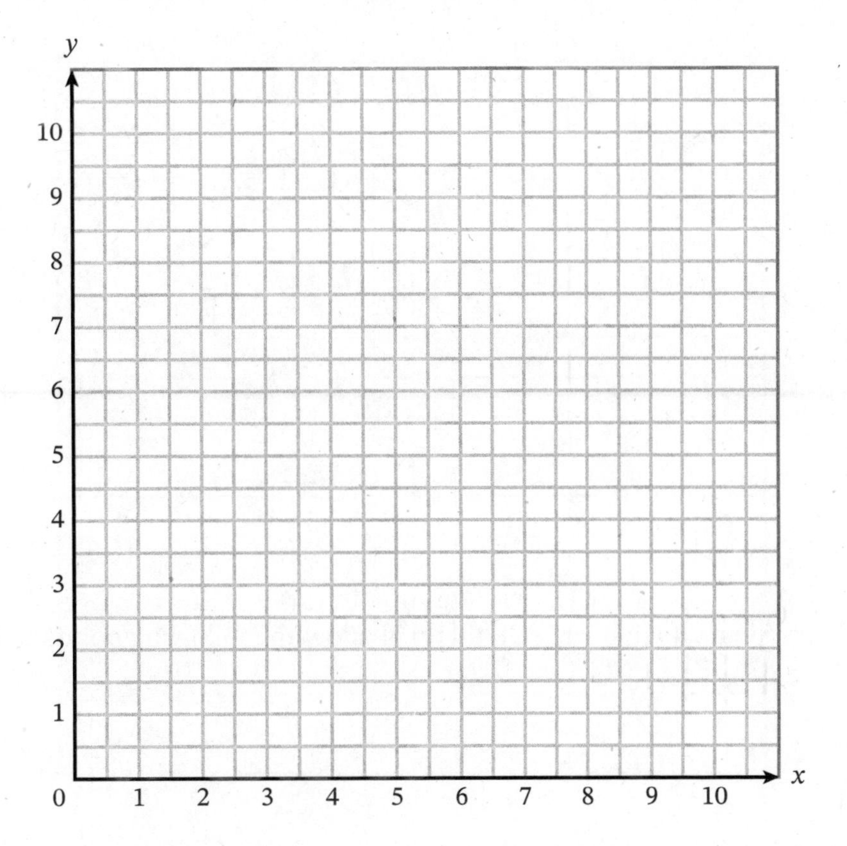

a. Plot and label the following points.

Point E (0, 4) Point F (4, 0) Point H $\left(\frac{1}{2}, 2\frac{1}{2}\right)$ Point I $\left(2\frac{3}{4}, 5\right)$

b. Point H is ______ units above the x-axis.

c. Point H is ______ units to the right of the y-axis.

d. The interval length of the x-axis is ______ units.

e. The interval length of the y-axis is ______ units.

Use the grid to complete problems 4 and 5.

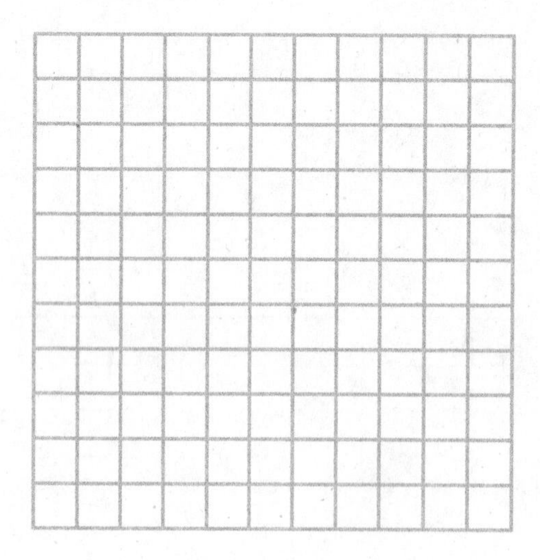

4. Draw a coordinate plane. Include a scale that will allow the following points to be plotted. Plot and label the points.

   Point J (18, 10)

   Point L (4, 11)

   Point M (18, 6)

5. Describe the similarities and differences between the locations of points M and J.

4

Name ____________________ Date __________

1. Use the coordinate plane to answer each question.

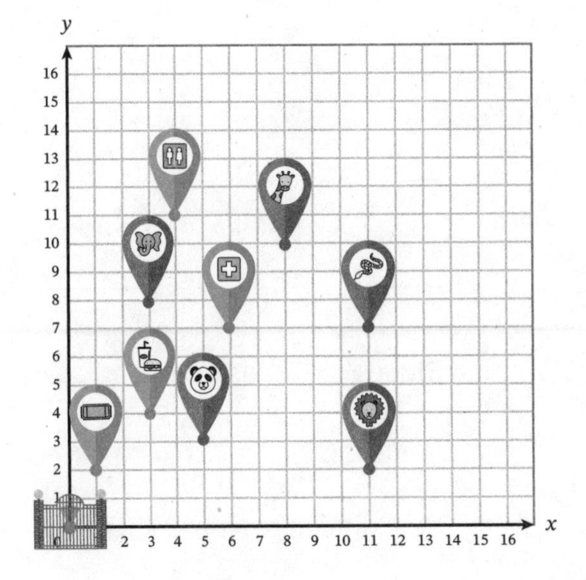

a. Kayla is at the ticket stand and wants to see the lions. How many units away from the ticket stand is the lion exhibit and in which direction?

12 unit to the right

b. Lacy eats lunch at the snack bar. Which animal exhibit is 4 units away from Lacy?

elephant

c. Which exhibit is 4 units to the right and 1 unit down from the restrooms?

Name Jayden Jr. Date

1. The graph shows the screen of a video game in which a rabbit must hop to various garden locations to find treats. Use this graph to complete parts (a)–(g).

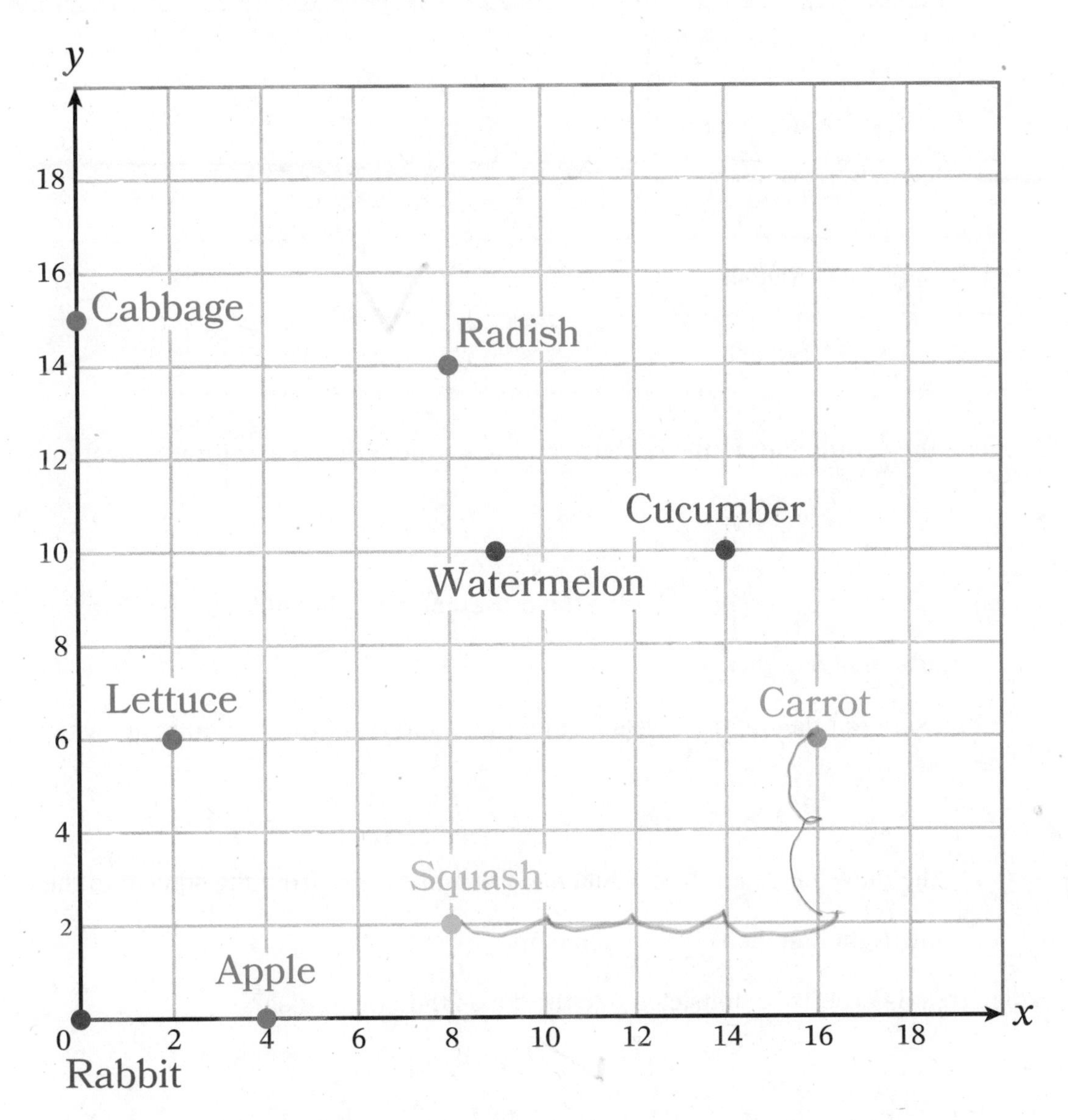

a. Complete the table with the ordered pair for each treat.

Treat	Ordered Pair
Apple	(4, 0)
Carrot	(16, 6)
Cucumber	(14, 10)
Lettuce	(2, 6)
Radish	(8, 14)
Squash	(8, 2)

b. Estimate the coordinates of the cabbage and write the ordered pair for the location of the cabbage.

(0, 15)

c. If a rabbit starts at the origin, describe the movements he should make to get to the cabbage.

__0__ units right and then __15__ units up

d. If a rabbit starts at the origin, travels 8 units right, and then travels 2 units up, which treat will he find?

Squash

e. Describe the movements a rabbit would need to make to get from the squash to the carrot.

__4__ units right and then __2__ units up

f. Which treat is exactly 12 units closer to the x-axis than the radish?

Squash

g. The farmer drops a banana on the ground. The banana is located closer to the x-axis than the lettuce and has the same x-coordinate as the apple. Write one possible ordered pair to represent the location of the banana.

(2, 4)

2. The coordinate plane shows the locations of landmarks in Atlanta, Georgia. Each unit represents 1 mile. Use this graph to complete parts (a)–(d).

a. Write the ordered pair for each location.

Location	Ordered Pair
Morehouse College	(0, 0)
Waterworks Park	
Oakland Cemetery	
Martin Luther King Jr.'s Birthplace	
Georgia Institute of Technology	

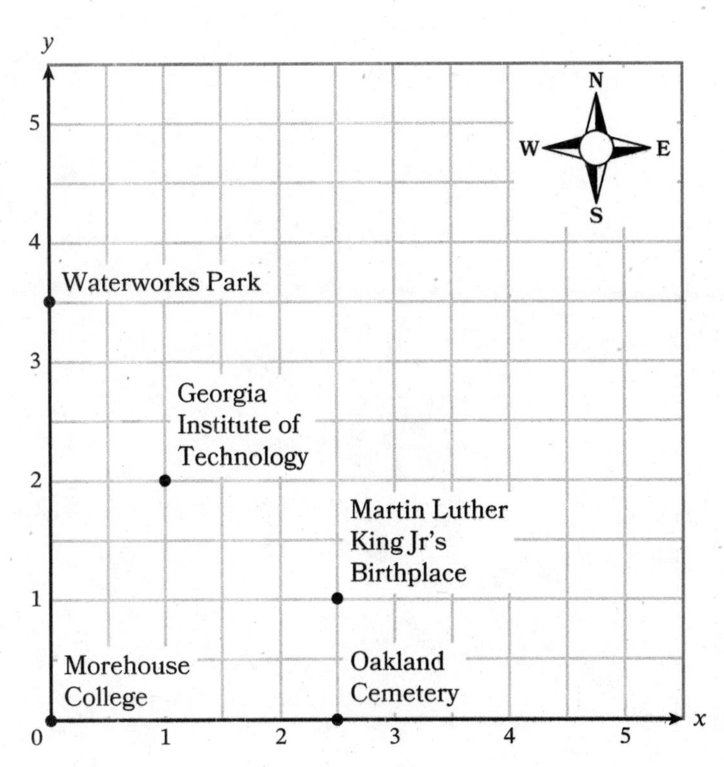

b. Which landmark is directly north of Oakland Cemetery?

c. How are the coordinates of the Oakland Cemetery and the coordinates of Martin Luther King Jr.'s birthplace the same and different?

d. Describe how to get from Martin Luther King Jr.'s birthplace to the Georgia Institute of Technology by moving horizontally and then vertically.

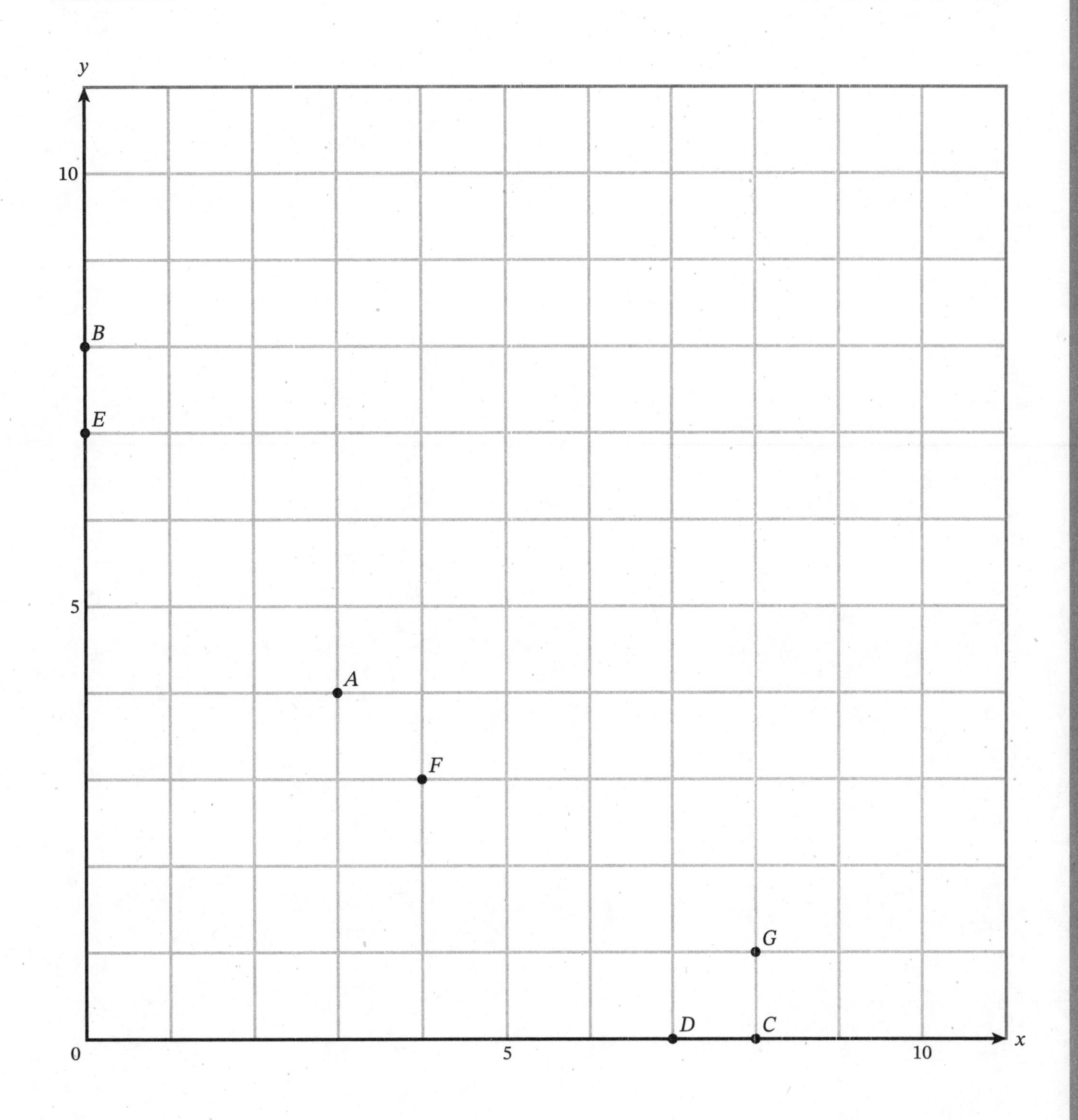
y
10
B
E
5
A
F
G
D
C
0
5
10
x

Version 1

Part A

Use the grid to construct a labeled coordinate plane. Then plot the points listed.

Ordered Pairs: (4, 7), (4, 10), (4, 0), (4, 1), (4, 4)

Part B

Write the ordered pairs for three points that lie on a horizontal line. Use different points from the ones listed above.

Write the ordered pairs for three points that lie on a vertical line. Use different points from the ones listed above.

Version 2

Part A

Use the grid to construct a labeled coordinate plane and plot the points listed.

Ordered Pairs: (2, 7), (4, 7), (0, 7), (10, 7), (9, 7)

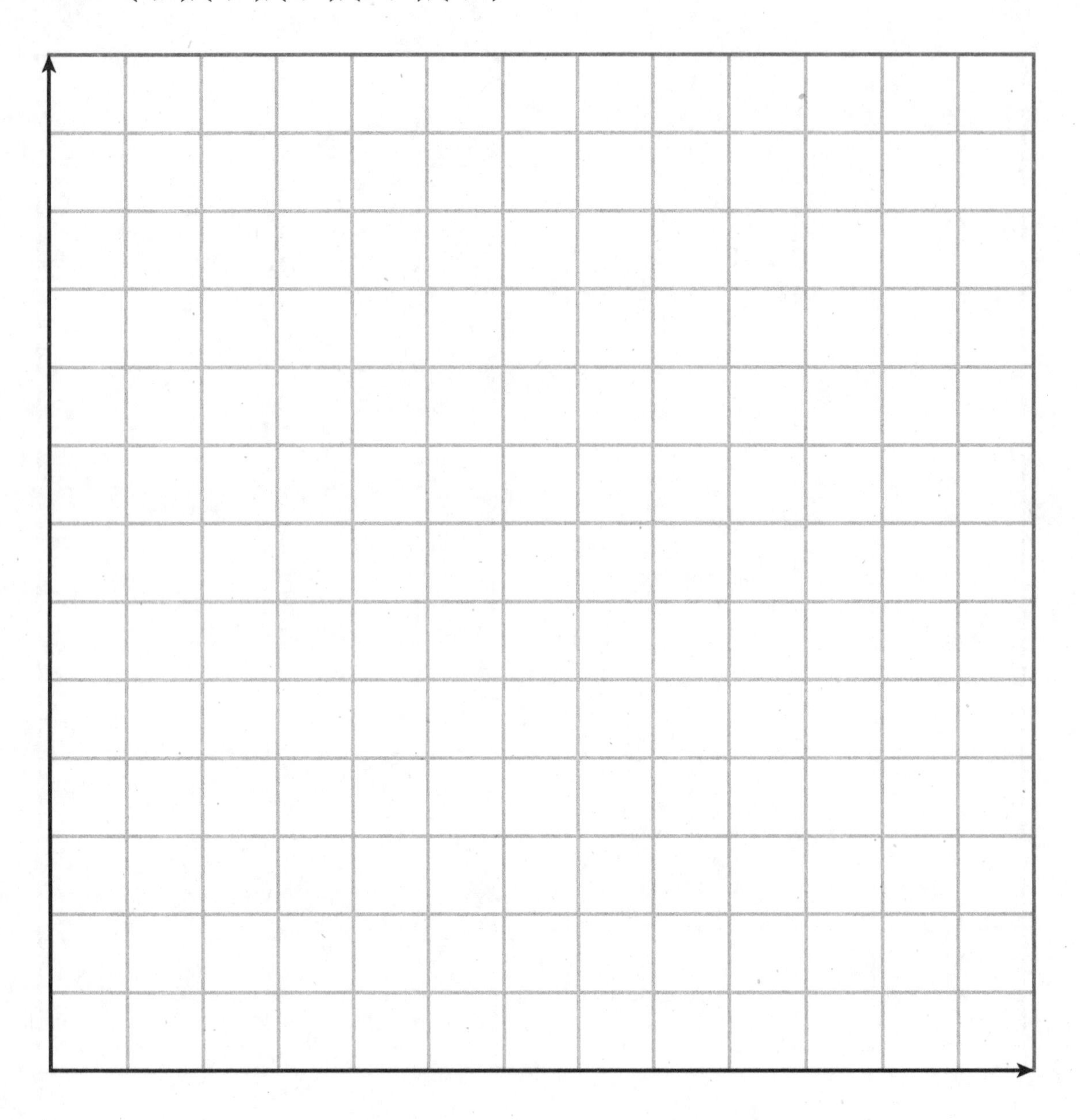

Part B

Write the ordered pairs for three points that lie on a horizontal line. Use different points from the ones listed above.

Write the ordered pairs for three points that lie on a vertical line. Use different points from the ones listed above.

5

Name ____________________ Date ________

1. Line ℓ is shown in the coordinate plane.

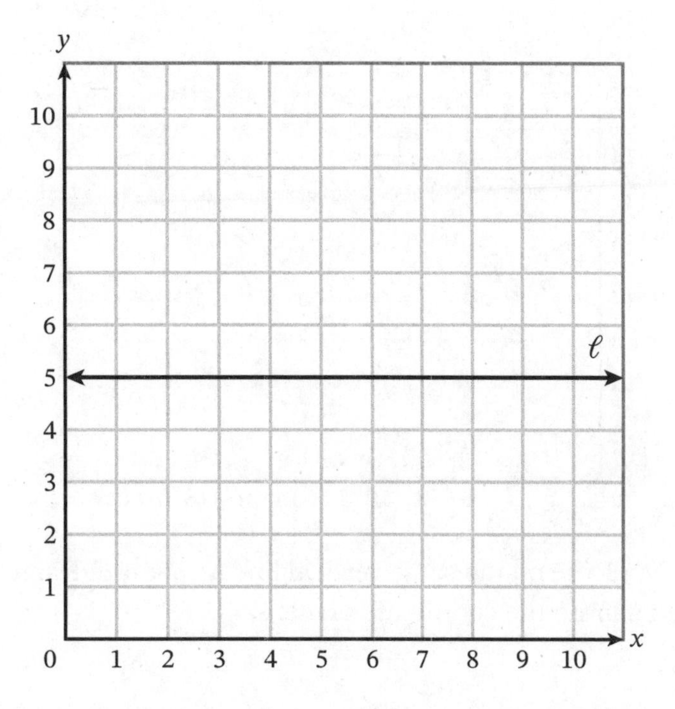

a. Draw a line in the coordinate plane that is parallel to the x-axis but a greater distance from the x-axis than line ℓ. Label this line m. Write the ordered pairs for three points on line m.

b. Draw a line in the coordinate plane that is parallel to the x-axis but a shorter distance from the x-axis than line ℓ. Label this line n. Write the ordered pairs for three points on line n.

2. The graph shows point F.

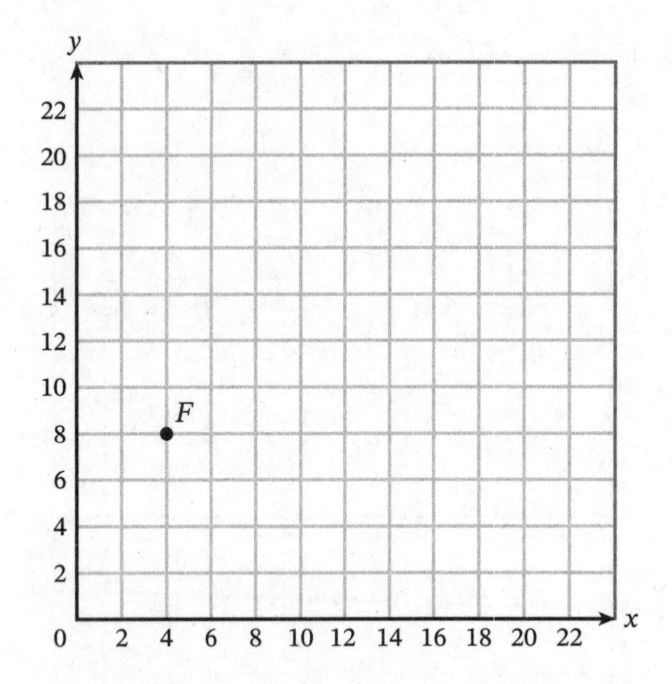

a. Plot a point that would lie on the same vertical line as point F. Name the new point E and record its ordered pair on the coordinate plane.

b. Draw $\overleftrightarrow{EF}$.

5

Name _______________________ Date ___________

1. Use the graph to complete parts (a)–(e).

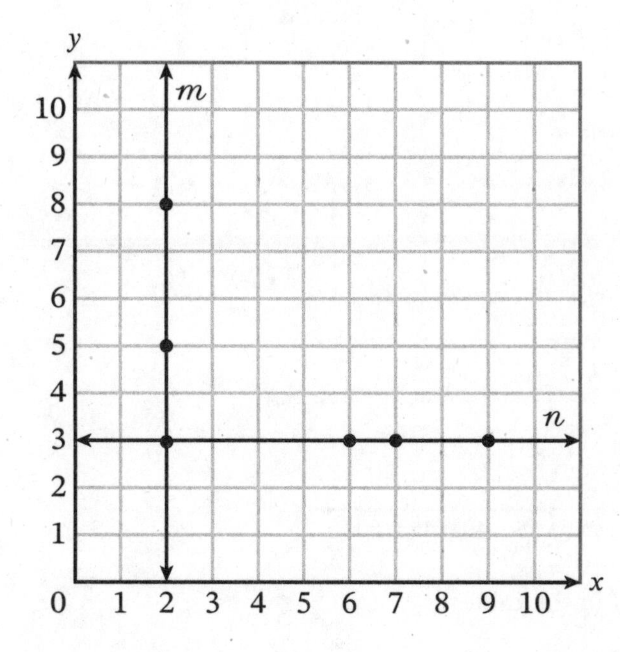

Points on Line m	Points on Line n
$(2, 3)$	$(6, 3)$
$(2, 5)$	$(7, 3)$
$(2, 8)$	$(9, 3)$

a. Highlight the coordinates that are the same for line m.

b. Is line m horizontal or vertical?

c. Highlight the coordinates that are the same for line n.

d. Is line n horizontal or vertical?

e. Line n is parallel to the ______-axis and perpendicular to the ______-axis.

2. Use the graph to complete parts (a)–(f).

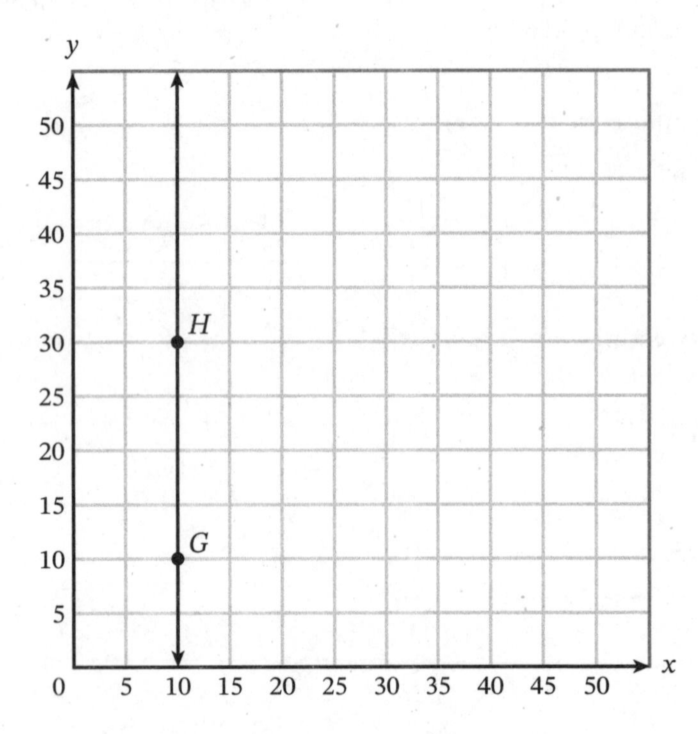

a. Write the ordered pairs for points G and H in the table.

Point	Ordered Pair
G	
H	

b. Write the ordered pair for a different point that also lies on $\overleftrightarrow{GH}$.

c. Is $\overleftrightarrow{GH}$ horizontal or vertical?

d. Draw a line in the coordinate plane that is parallel to the y-axis but is a greater distance from the y-axis than $\overleftrightarrow{GH}$. Label this line p. Write the ordered pairs for three points on the line.

e. Draw a line in the coordinate plane that is parallel to the y-axis but a shorter distance from the y-axis than $\overleftrightarrow{GH}$. Label this line ℓ.

f. Write the ordered pairs for three points on line ℓ.

---

3. The ordered pairs for four points are shown. Do the points lie on a horizontal line, a vertical line, or neither? How do you know?

$(45, 2)$

$(45, 60)$

$(45, 15)$

$(45, 34)$

---

4. The ordered pairs for four points are shown. Do the points lie on a horizontal line, a vertical line, or neither? How do you know?

$(4, 0)$

$(0, 0)$

$(0, 9)$

$\left(39\frac{1}{2}, 0\right)$

5. The points (2, 5) and (7, 5) lie on line a.

   a. Is line a horizontal or vertical?

   b. Line a is parallel to the ______-axis.

   c. Line a is perpendicular to the ______-axis.

   d. Write the ordered pair for another point that lies on line a.

   e. Describe the distance between line a and the x-axis. Explain how you know.

5

Name ______________________ Date ____________

Use the graph to complete parts (a)–(d).

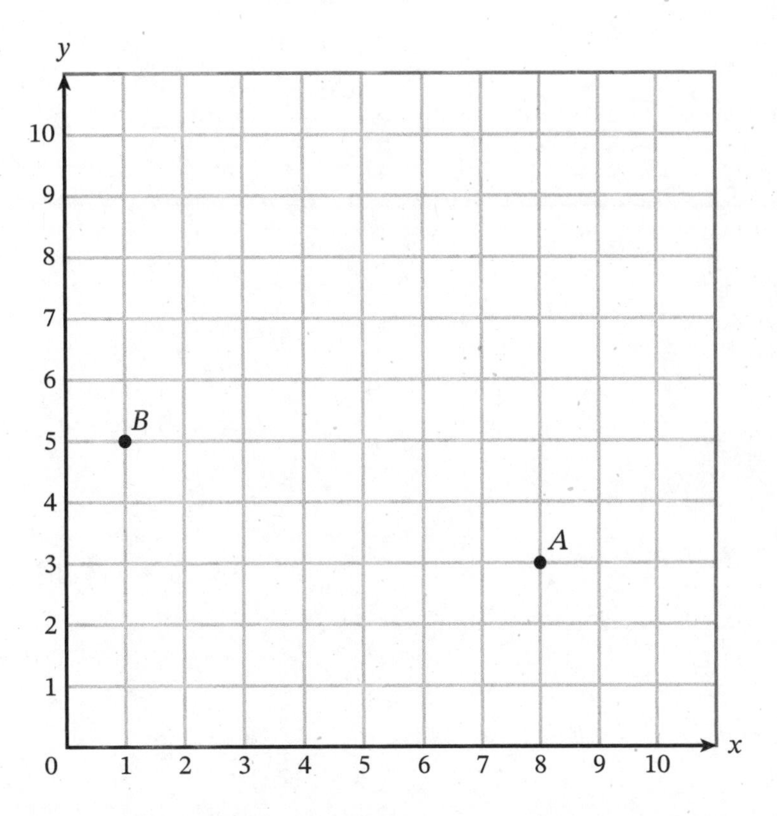

a. Plot a point that lies on the same vertical line as point B. Name the new point C. Record its ordered pair beside it on the coordinate plane.

b. What do the coordinates for point B and point C have in common?

c. Plot a point that lies on the same horizontal line as point A. Name the new point D. Record its ordered pair beside it on the coordinate plane.

d. What do the coordinates for point A and point D have in common?

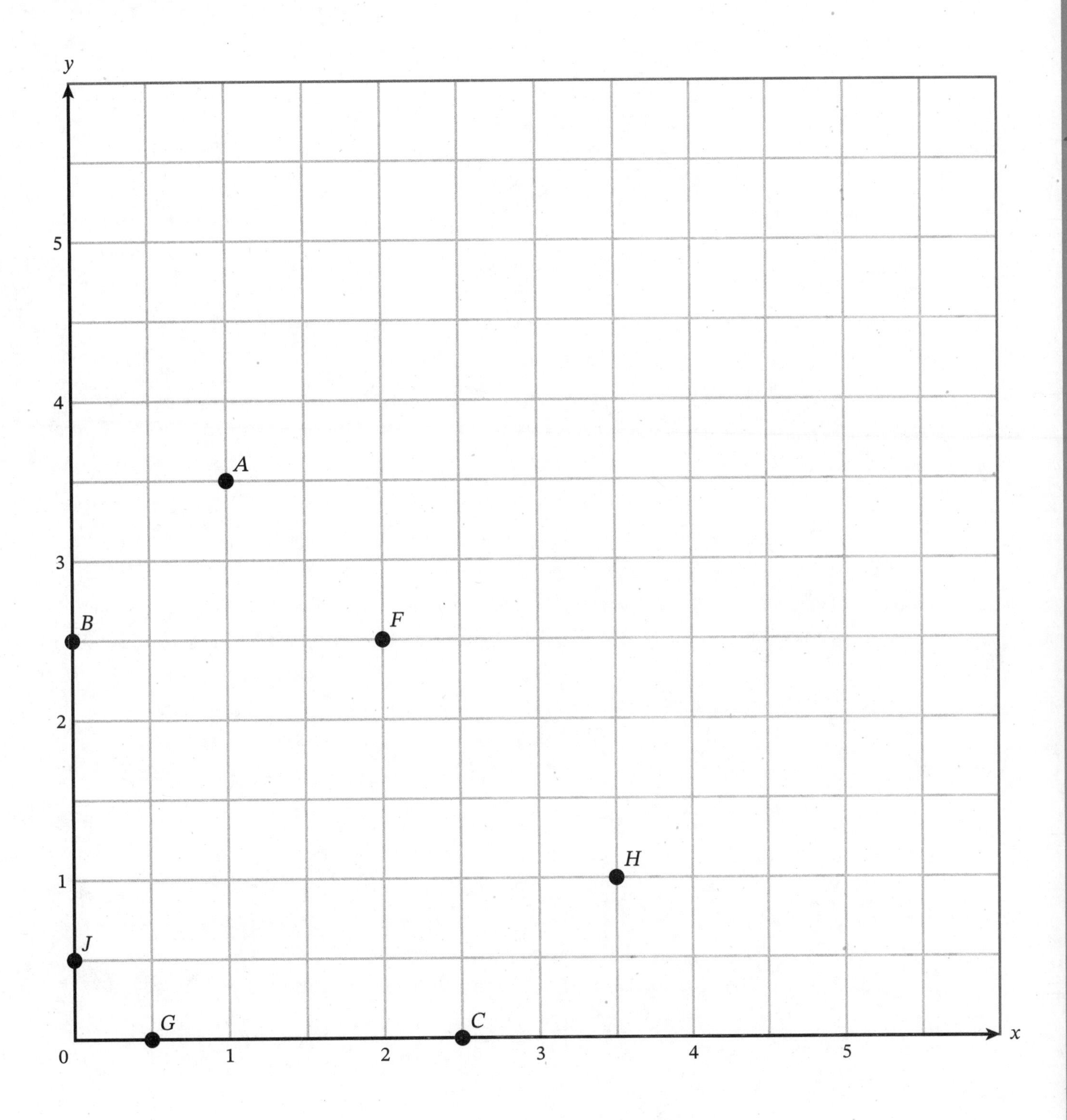
y
5
4
3
2
1
0
1
2
3
4
5
x
A
B
F
H
J
G
C

6

Name ______________________ Date __________

1. Use the coordinate plane to complete parts (a)–(h).

   a. Draw and label point A at $(8, 1)$.

   b. Draw a line that is perpendicular to the x-axis through point A. Label the line m.

   c. Plot a point on line m that is 6 units farther from the x-axis than point A. Label this point B and write its ordered pair next to it.

   d. Plot a point on line m that is halfway between points A and B. Label this point C and write its ordered pair next to it.

   e. Draw line n so that it is 2 units from the x-axis and $n \perp m$.

   f. Point E is on line n. It is 2 units from the y-axis. Plot point E and write its ordered pair next to it.

   g. Draw line ℓ so that it passes through point E and $\ell \parallel m$.

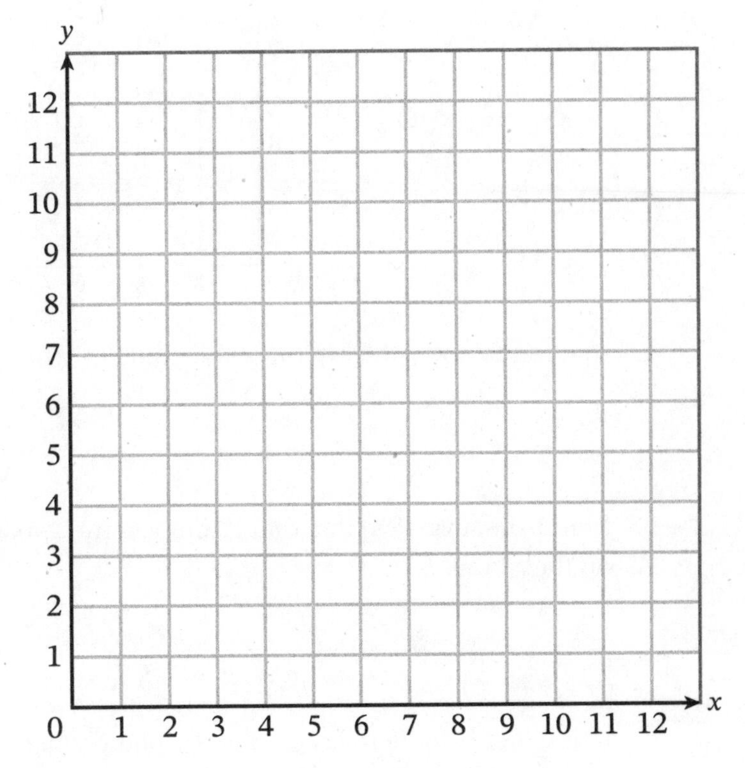

   h. Point F is on line ℓ and is farther from the x-axis than point E. Plot point F and write its ordered pair next to it.

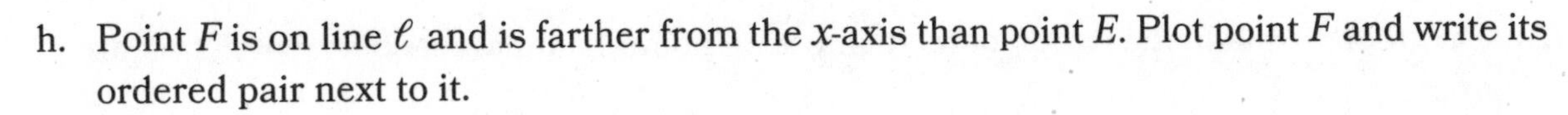

2. The graph shows line ℓ.

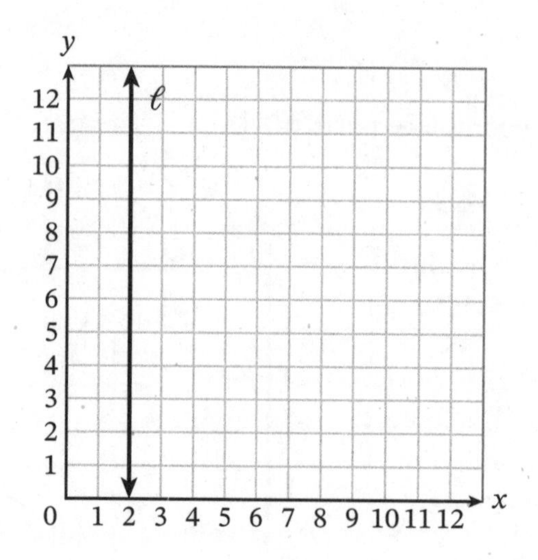

a. Use red to color the region of the plane in which all x-coordinates are less than 2.

b. Line m is a vertical line that intersects the x-axis at the point (8, 0). Draw and label line m on the graph.

c. Use blue to shade the region of the plane where points are more than 8 units from the y-axis.

d. Use green to shade the region of the plane with points that have x-coordinates that are greater than 2 and less than 8.

6

Name ______________________ Date ____________

1. Indicate whether each statement about $\overleftrightarrow{DE}$ is true or false.

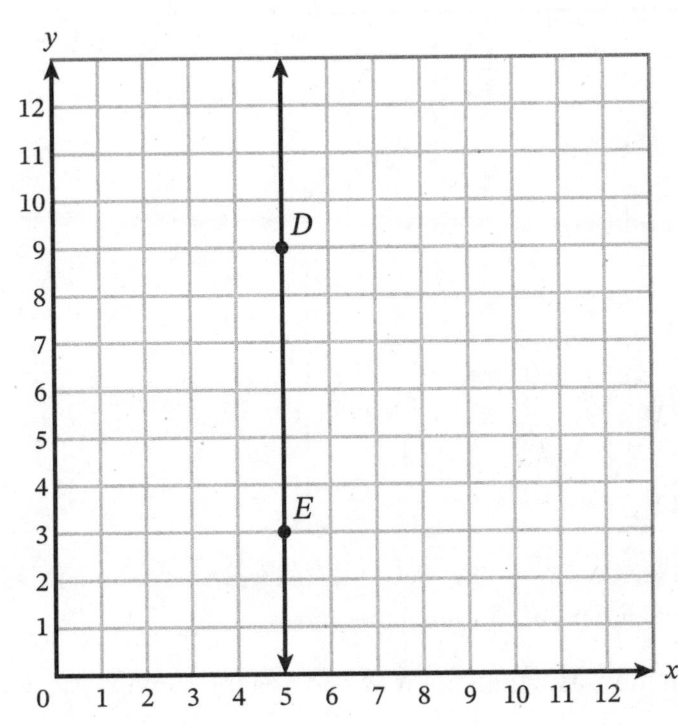

Statement	True	False
Each point on $\overleftrightarrow{DE}$ has the same x-coordinate.		
Each point on $\overleftrightarrow{DE}$ has the same y-coordinate.		
All points on $\overleftrightarrow{DE}$ are collinear.		
Each point on $\overleftrightarrow{DE}$ is the same distance from the x-axis.		
Each point on $\overleftrightarrow{DE}$ is the same distance from the y-axis.		
$\overleftrightarrow{DE}$ is horizontal.		
$\overleftrightarrow{DE}$ is vertical.		

2. Use the graph to complete parts (a)–(d).

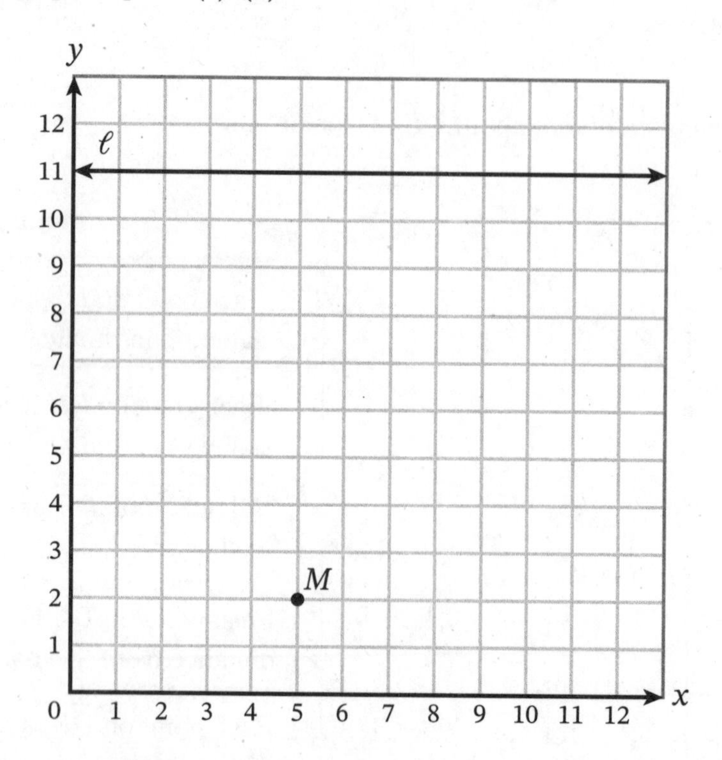

a. Draw a line that is parallel to line ℓ through point M. Label the line p.

b. Plot a point that is 4 units farther from the x-axis than point M and has the same x-coordinate as point M. Label the point N.

c. Draw a line through points M and N. Label the line c.

d. Fill in each blank with ∥ or ⊥ to make the statement true.

ℓ ______ p

ℓ ______ c

c ______ p

3. Use the coordinate plane to complete parts (a)–(d).

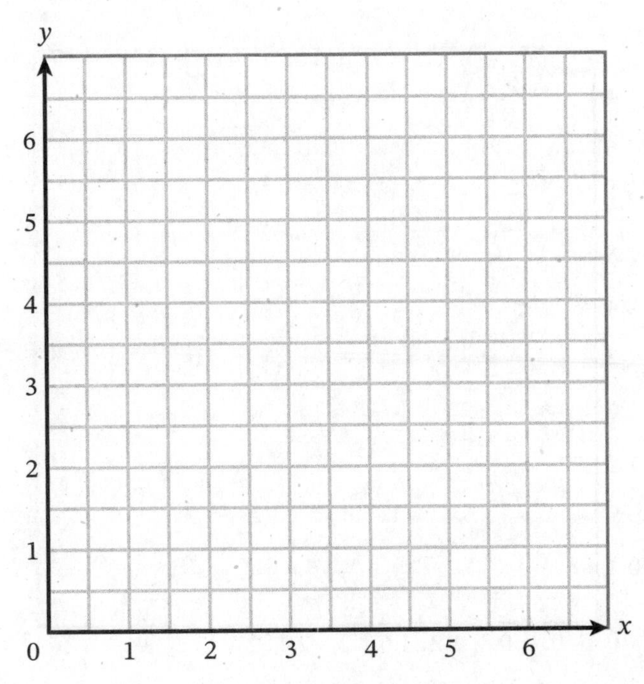

a. Draw a line, n, that is parallel to the y-axis and $4\frac{1}{2}$ units from the y-axis.

b. Write the ordered pair for the point on line n that is 4 units from the x-axis.

c. Lightly shade the region of the plane where points are greater than $4\frac{1}{2}$ units from the y-axis.

d. Circle the ordered pairs for points that appear in the shaded region.

$\left(5, 4\frac{1}{2}\right)$ $\left(4\frac{3}{8}, 5\right)$ $\left(4\frac{5}{8}, 5\right)$ (4, 3.5) (12, 3)

4. Use the coordinate plane to complete parts (a)–(e).

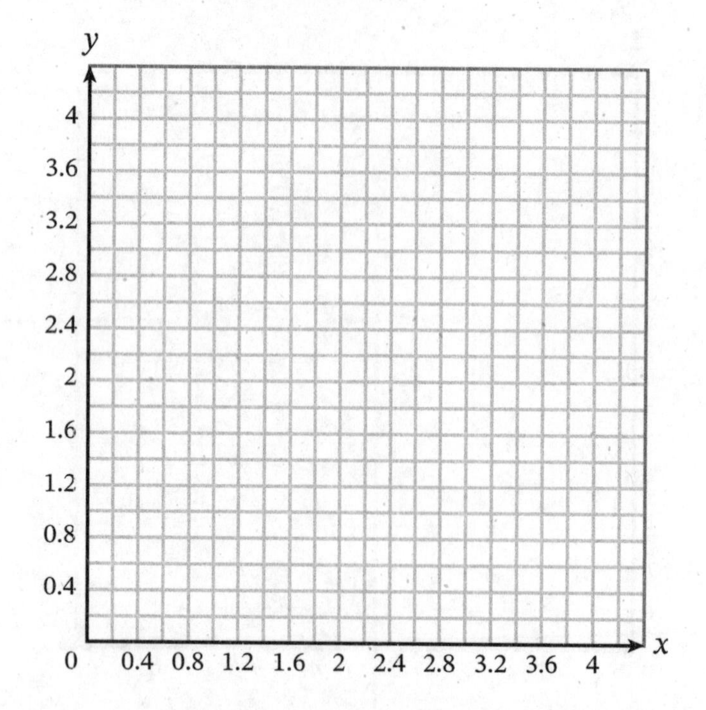

a. Draw a line, a, that is parallel to the x-axis and 0.8 units from the x-axis.

b. Draw a line, d, that passes through the point $(1.2, 2.4)$ and $d \parallel a$.

c. Lightly shade the region between line d and line a. Complete the statement.

All points in the shaded region have a y-coordinate greater than ______ but less than ______ .

d. Is point $R(6, 1.8)$ in the shaded region? Explain.

e. What is the distance between line a and line d?

6

Name ____________________ Date __________

1. Use the graph to complete parts (a)–(d).

a. Draw a vertical line through point A.

b. Plot a point on the vertical line that is 2 units farther from the x-axis than point A. Label this point B and write the ordered pair next to it.

c. Is $\overleftrightarrow{AB}$ parallel or perpendicular to the y-axis?

d. Is $\overleftrightarrow{AB}$ parallel or perpendicular to line m?

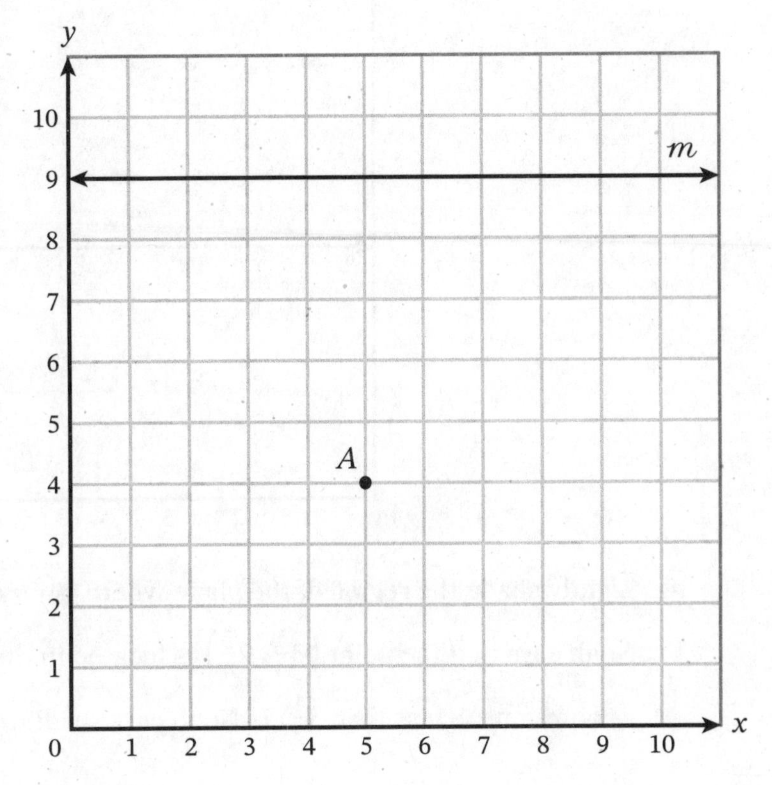

2. Use the graph of line n to complete parts (a)–(b).

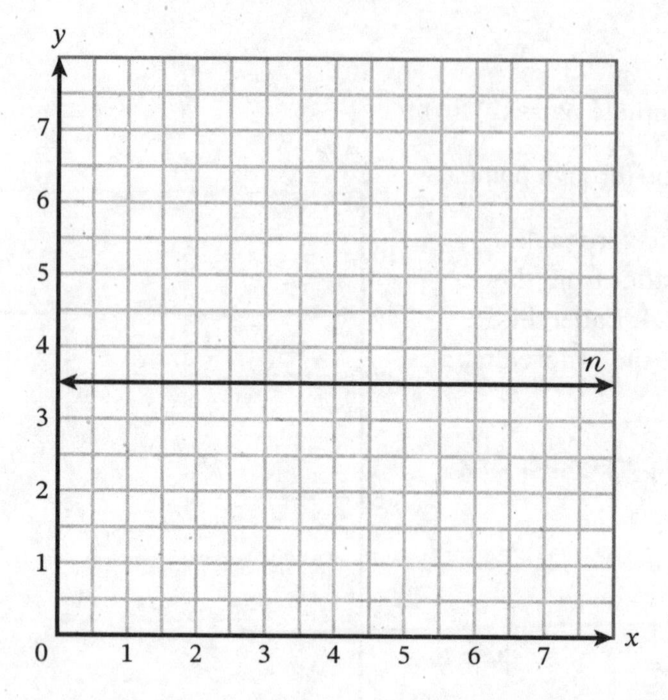

a. Lightly shade the region of the plane where points are less than $3\frac{1}{2}$ units from the x-axis.

b. Noah says that the point $\left(3\frac{1}{4}, 2\frac{3}{4}\right)$ is located in the shaded region because the y-coordinate is less than $3\frac{1}{2}$. Is Noah correct? Explain.

7

Name ______________________ Date __________

1. The table shows the first three terms in pattern A and in pattern B.

Pattern A	0	3	6		
Pattern B	0	$2\frac{1}{2}$	5		

a. Complete each pattern in the table.

b. What will be the number in pattern B when the number in pattern A is 18?

c. What will be the number in pattern A when the number in pattern B is $17\frac{1}{2}$?

2. Leo and Sasha create number patterns.

Leo's pattern: Start at 6 and multiply by 4.

Sasha's pattern: Start at 85 and subtract 6.

Record the first five terms of Leo's pattern and of Sasha's pattern in the table.

Leo's Pattern					
Sasha's Pattern					

3. Use the table to complete parts (a)–(c).

   a. Use the rules to complete the patterns.

   b. Write the ordered pair for each pair of corresponding terms by writing the number from pattern A as the x-coordinate and the number from pattern B as the y-coordinate.

Pattern A Add 2	**Pattern B** Add 3	**Ordered Pair**
0	0	
2	3	
4	6	

   c. Plot the points in the coordinate plane.

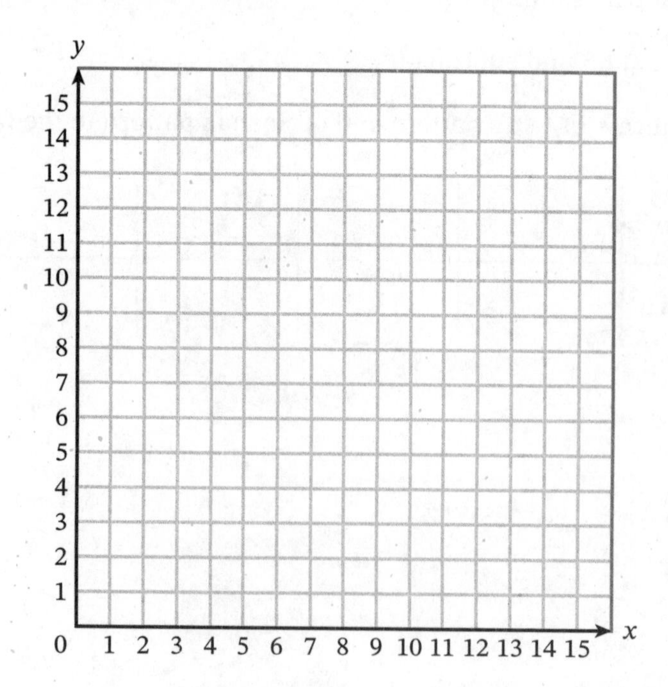

7

Name ______________________ Date __________

1. Use pattern N to complete parts (a) and (b).

Pattern N	0	$\frac{1}{2}$	1		2	

a. Write the rule for pattern N.

b. Complete the table.

2. Use the table to complete parts (a)–(c).

a. The rule for pattern Y is add 4. The rule for pattern Z is subtract $\frac{1}{2}$. Complete the table.

Pattern Y	Pattern Z
0	4
4	$3\frac{1}{2}$
8	3

b. What is the number in pattern Z when the number in pattern Y is 24?

c. What is the number in pattern Y when the number in pattern Z is 0?

3. Use the table of ordered pairs to complete parts (a)–(e).

a. Write the rule for pattern A.

b. Write the rule for pattern B.

c. Use the numbers from pattern A and pattern B to create ordered pairs and complete the table.

d. Plot the ordered pairs from the table in the coordinate plane.

Pattern A x	Pattern B y	Ordered Pair (x, y)
0	4	
3	5	
6	6	
9	7	

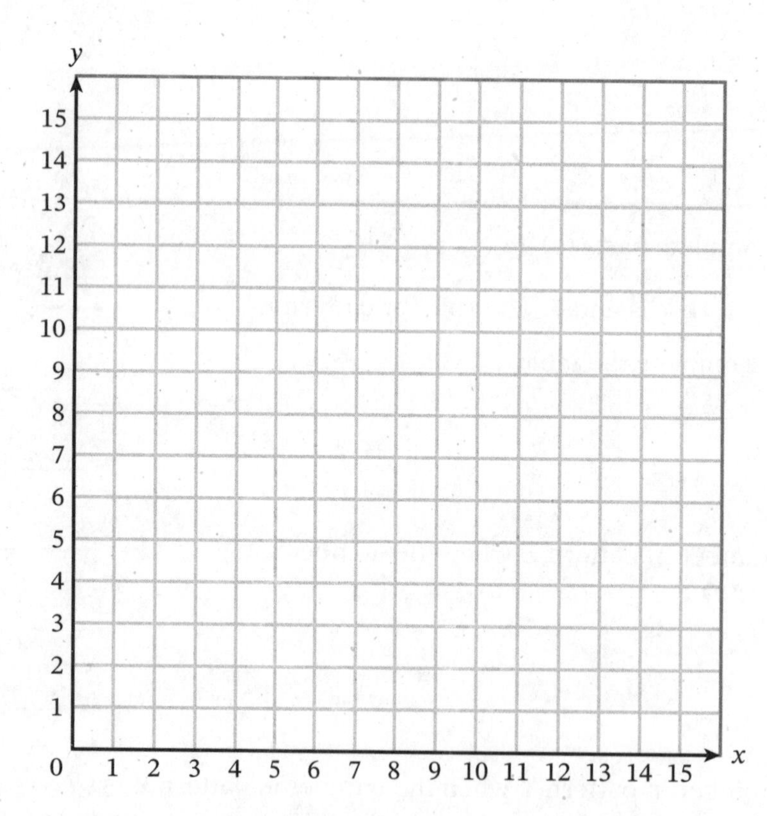

e. Describe the movement needed to get from point (6, 6) to (9, 7).

4. Use the table to complete parts (a)–(c).

Pattern P x	**Pattern Q** y	**Ordered Pair** (x, y)
2	0	$(2, 0)$
$3\frac{1}{2}$	3	$\left(3\frac{1}{2}, 3\right)$
5	6	$(5, 6)$
$6\frac{1}{2}$	9	$\left(6\frac{1}{2}, 9\right)$

a. Every time a number in pattern P increases by $1\frac{1}{2}$, what happens to the numbers in pattern Q?

b. If patterns P and Q continue, what will be the next ordered pair on the table?

c. If patterns P and Q continue, would the ordered pair $(11, 18)$ be part of the pattern? How do you know?

7

Name _______________ Date _______________

Consider the table shown.

Pattern A	3					
Pattern B	6					

The rule for pattern A is add $1\frac{1}{2}$.

The rule for pattern B is add 3.

a. Complete the table by using the rules for pattern A and pattern B.

b. What is the number in pattern B when the number in pattern A is 15?

8

Name ____________________ Date __________

1. The rule for the x-coordinate is add 1.5. The rule for the y-coordinate is add 1.5.

   a. Complete the table.

x-Coordinate	y-Coordinate	Ordered Pair
1	5	(1, 5)

   b. Plot the four ordered pairs in the coordinate plane.

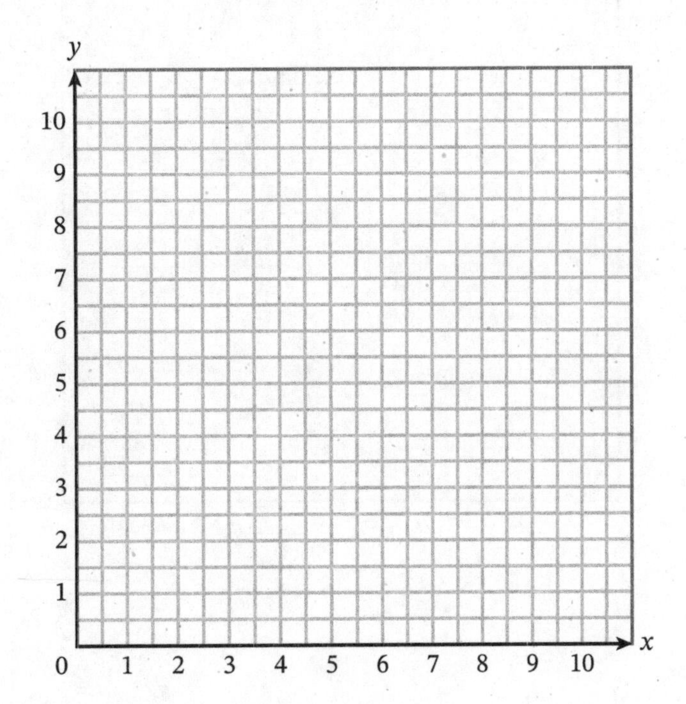

2. Each y-coordinate is 1.5 more than its corresponding x-coordinate.

   a. Complete the table.

x-Coordinate	Calculation	y-Coordinate	Ordered Pair
0	$0 + 1.5 = 1.5$	1.5	(0, 1.5)
3			
6			
9			

   b. Plot the four ordered pairs in the coordinate plane.

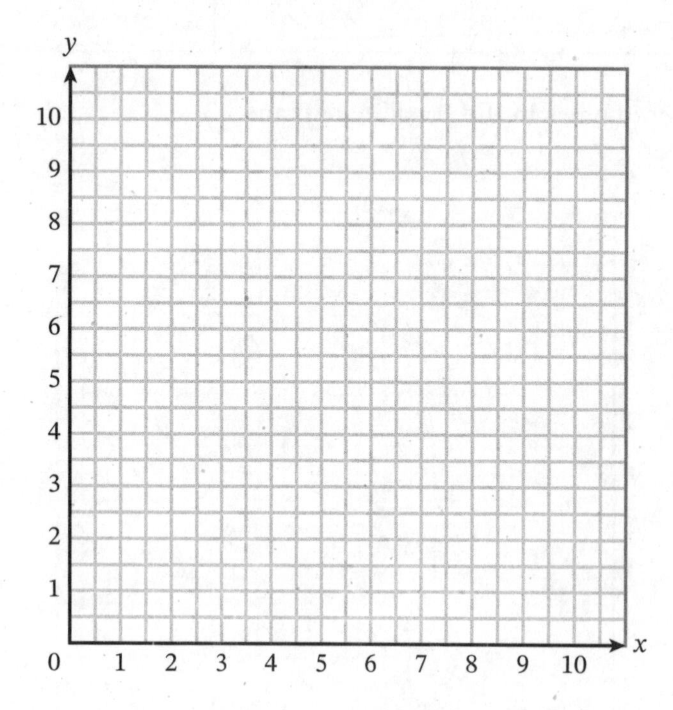

3. Use the graph to complete parts (a)–(f).

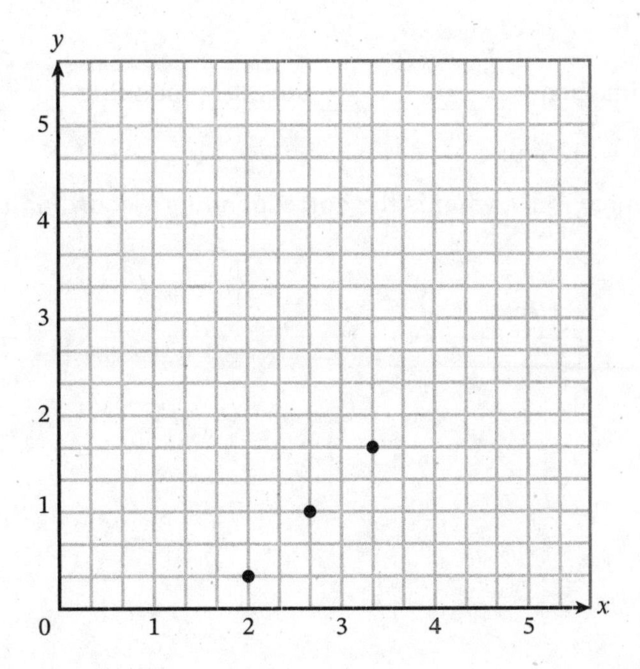

a. Describe the movement from one point to the next.

b. What is the rule for the x-coordinate?

c. What is the rule for the y-coordinate?

d. Use the rules for the coordinates to plot the next three points in the coordinate plane. What are the ordered pairs for the points?

e. Fill in the blanks to describe the relationship between the x- and y-coordinates.

The ______-coordinates are ____________ the corresponding ______-coordinates.

f. When the x-coordinate is 10, what is the corresponding y-coordinate? Show how you know.

8

Name ______________________ Date __________

1. Use the table and graph to complete parts (a)–(c).
   - Rule for the x-coordinate: Add 4
   - Rule for the y-coordinate: Add 4

x-Coordinate	y-Coordinate	Ordered Pair
2	3	(2, 3)
6	7	(6, 7)
10	11	(10, 11)
14	15	(14, 15)

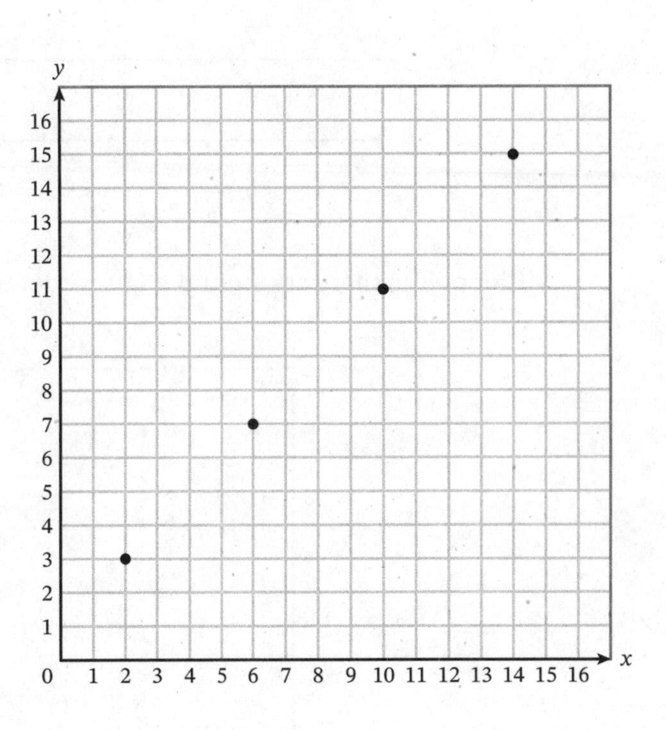

a. To get from point (2, 3) to point (6, 7), move right ______ units and then move up ______ units.

b. When the x-coordinate is 18, the corresponding y-coordinate is ______.

c. When the y-coordinate is 22, the corresponding x-coordinate is ______.

2. Use the table to complete parts (a)–(e).

x-Coordinate	y-Coordinate	Ordered Pair
0	2	(0, 2)
$1\frac{1}{2}$	$3\frac{1}{2}$	$\left(1\frac{1}{2}, 3\frac{1}{2}\right)$
3	5	(3, 5)
$4\frac{1}{2}$	$6\frac{1}{2}$	$\left(4\frac{1}{2}, 6\frac{1}{2}\right)$

a. Plot points that represent the four ordered pairs in the coordinate plane.

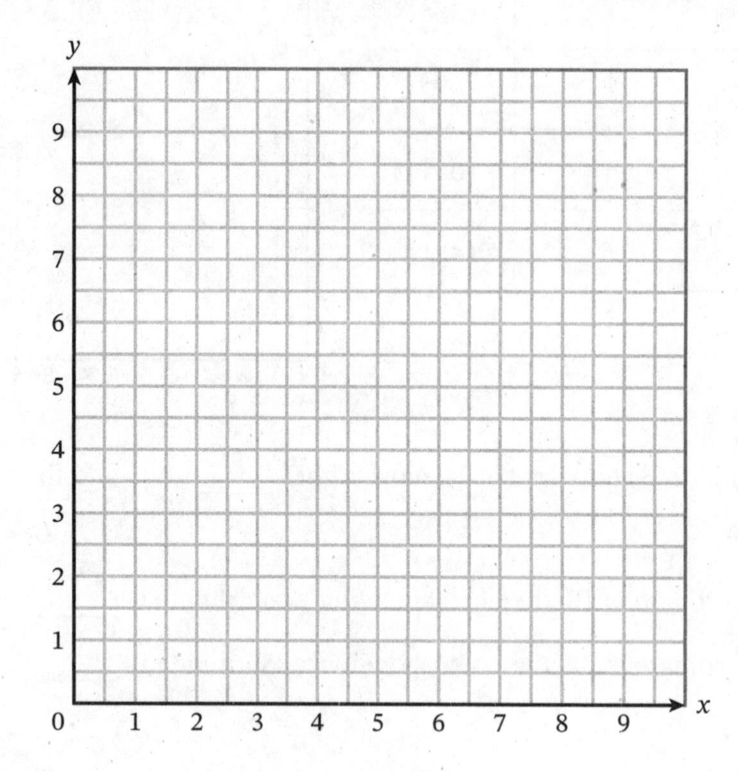

b. What is the rule for the x-coordinate?

c. What is the rule for the y-coordinate?

d. Describe the movement from point (3, 5) to point $\left(4\frac{1}{2}, 6\frac{1}{2}\right)$.

e. Fill in the blanks to describe the relationship between the x- and y-coordinates.

The ______-coordinates are 2 more than the corresponding ______-coordinates.

3. Use the graph to complete parts (a)–(g).

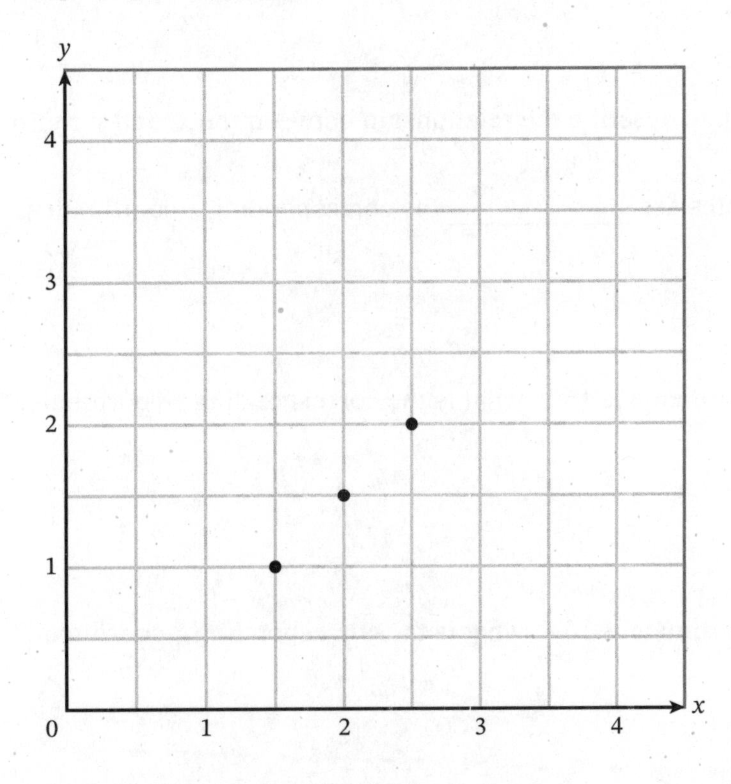

a. Describe the movement from one point to the next.

b. What is the rule for the x-coordinate?

c. What is the rule for the y-coordinate?

d. Use the rules for the coordinates to plot the next three points in the coordinate plane. What are the ordered pairs for the points?

e. Fill in the blank to describe the relationship between the x- and y-coordinates.

The y-coordinates are ____________ the corresponding x-coordinates.

f. When the x-coordinate is $16\frac{1}{2}$, what is the corresponding y-coordinate?

g. When the y-coordinate is $16\frac{1}{2}$, what is the corresponding x-coordinate?

8

Name ______________________ Date __________

Each y-coordinate is 5 more than its corresponding x-coordinate.

a. Complete the table.

x-Coordinate	Calculation	y-Coordinate	Ordered Pair
1	$1 + 5 = 6$	6	(1, 6)
3			
5			
7			

b. Plot the four ordered pairs in the coordinate plane.

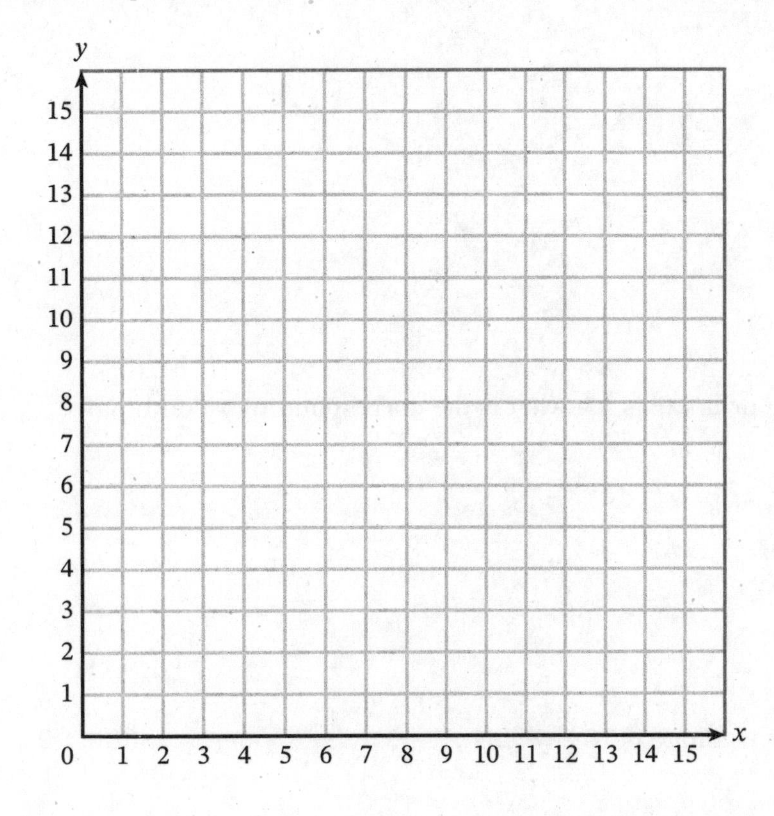

c. What is the rule for the x-coordinate?

d. What is the rule for the y-coordinate?

e. When the x-coordinate is 15, what is the corresponding y-coordinate?

9

Name ______________________ Date __________

1. Consider the coordinates and ordered pairs in the table.

   a. Complete the table.

x-Coordinate	*y*-Coordinate	Ordered Pair
1	2	(1, 2)
3	6	(3, 6)
5	10	(5, 10)

   b. Plot the six ordered pairs in the coordinate plane.

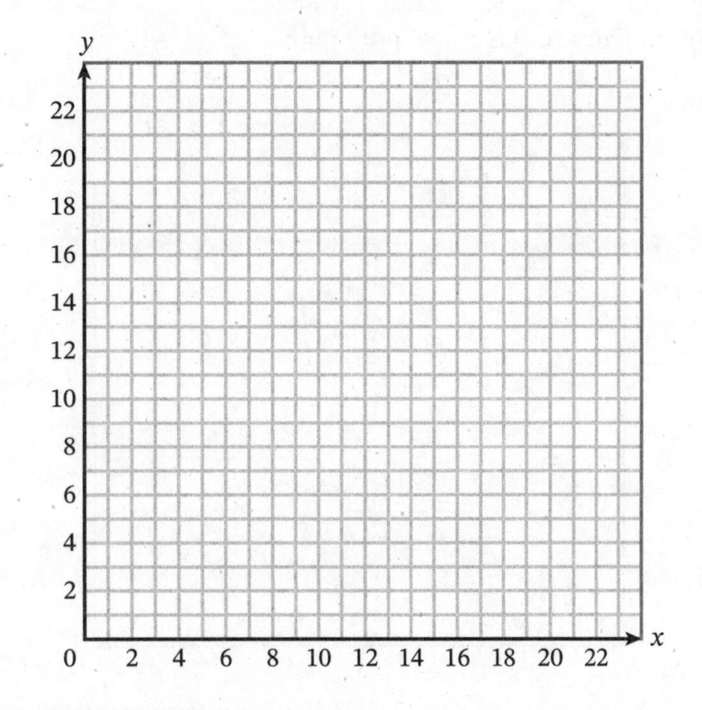

c. Describe the movement from one point to the next.

---

2. Multiply each x-coordinate by 4 to get its corresponding y-coordinate.

a. Complete the table.

x-Coordinate	Calculation	y-Coordinate	Ordered Pair
0	$0 \times 4 = 0$	0	(0, 0)
$\frac{1}{4}$			
$\frac{1}{2}$			
$\frac{3}{4}$			
1			

b. Plot the five ordered pairs in the coordinate plane.

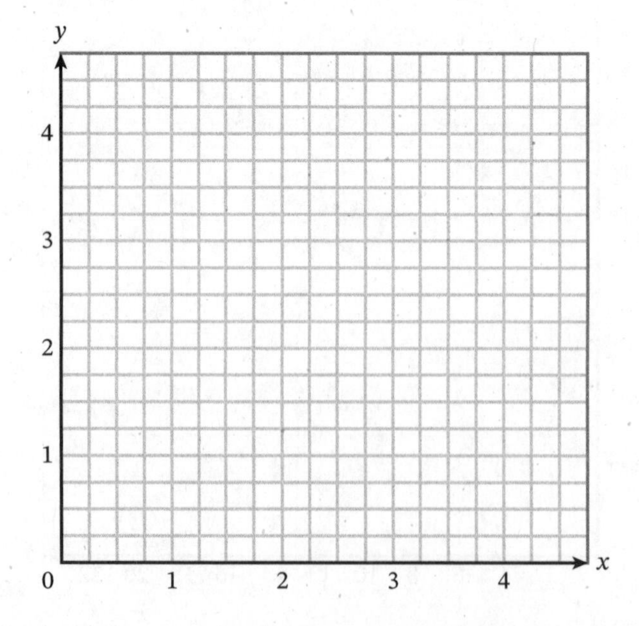

c. Describe the movement from one point to the next.

d. What is the rule for the x-coordinate?

e. What is the rule for the y-coordinate?

f. Fill in the blanks to describe the relationship between the x- and y-coordinates.

The ______-coordinates are ____

the corresponding ______-coordinates.

g. When the x-coordinate is $\frac{7}{2}$, what is the corresponding y-coordinate? Show how you know.

h. When the y-coordinate is 20, what is the corresponding x-coordinate? Show how you know.

3. Use the graph to complete parts (a)–(d).

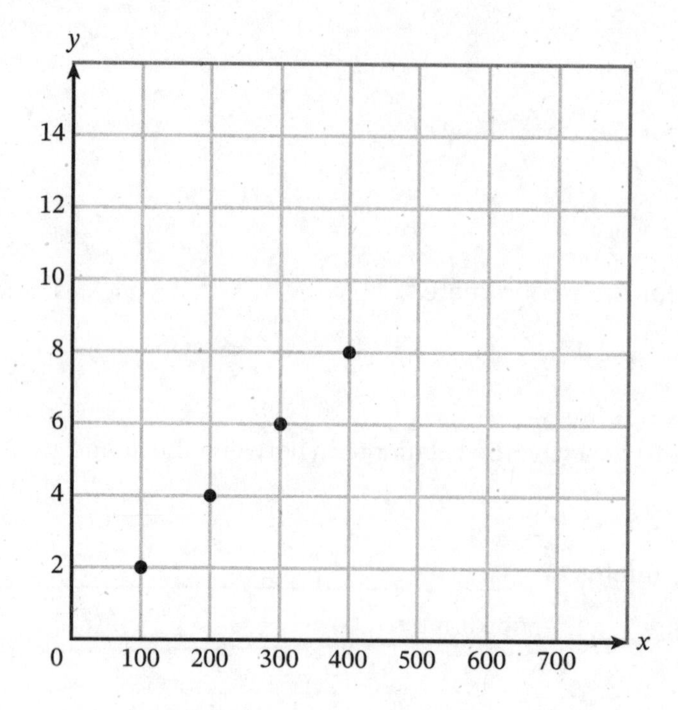

a. Use the rules for the coordinates to plot the next three points in the coordinate plane. What are the ordered pairs for the points?

b. Fill in the blanks to describe the relationship between the x- and y-coordinates.

The ______-coordinates are ____

the corresponding ______-coordinates.

c. What is the corresponding y-coordinate when the x-coordinate is 1,000? Show how you know.

d. What is the corresponding x-coordinate when the y-coordinate is 1,000? Show how you know.

9

Name ____________________ Date __________

1. Use the table and graph to complete parts (a)–(c).

Rule: Add 1 x-Coordinate	Rule: Add 3 y-Coordinate	Ordered Pair
0	0	(0, 0)
1	3	(1, 3)
2	6	(2, 6)
3	9	(3, 9)

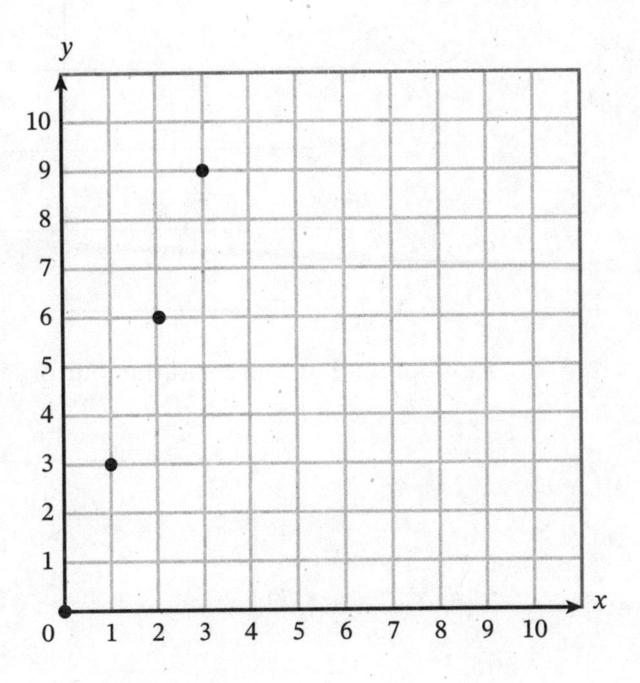

a. To get from point (0, 0) to point (1, 3), move right ______ units and then move up ______ units.

b. When the x-coordinate is 4, the corresponding y-coordinate is ______.

c. When the y-coordinate is 15, the corresponding x-coordinate is ______.

2. Use the table and coordinate plane to complete parts (a)–(e).

x-Coordinate	y-Coordinate	Ordered Pair
4	2	(4, 2)
8	4	(8, 4)
12	6	(12, 6)
16	8	(16, 8)

a. Plot points that represent the four ordered pairs in the coordinate plane.

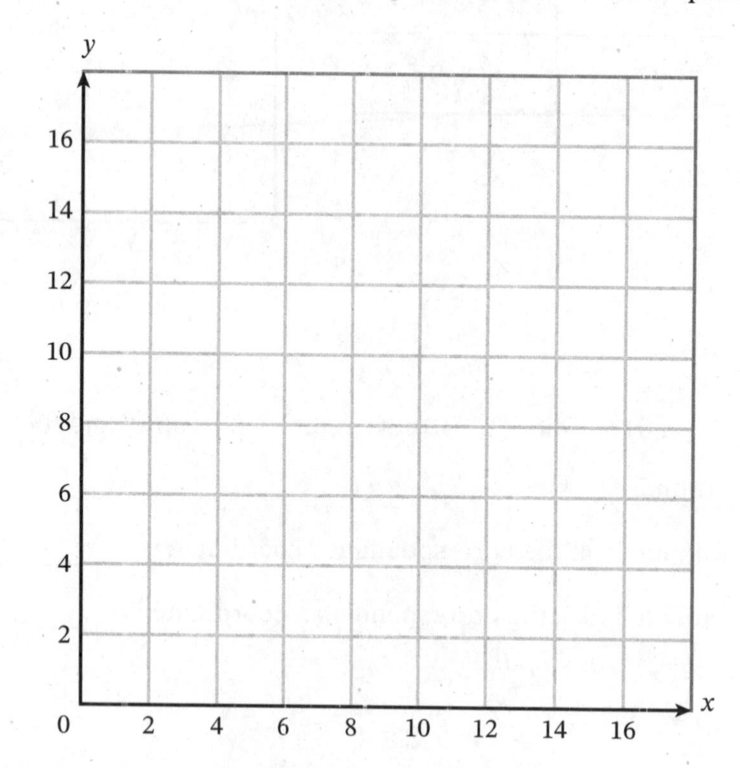

b. What is the rule for the x-coordinate?

c. What is the rule for the y-coordinate?

d. Describe the movement to get from point (4, 2) to point (8, 4).

e. Fill in the blanks to describe the relationship between the x- and y-coordinates.

The ______-coordinates are $\frac{1}{2}$ as much as the corresponding ______-coordinates.

---

3. Use the graph to complete parts (a)–(f).

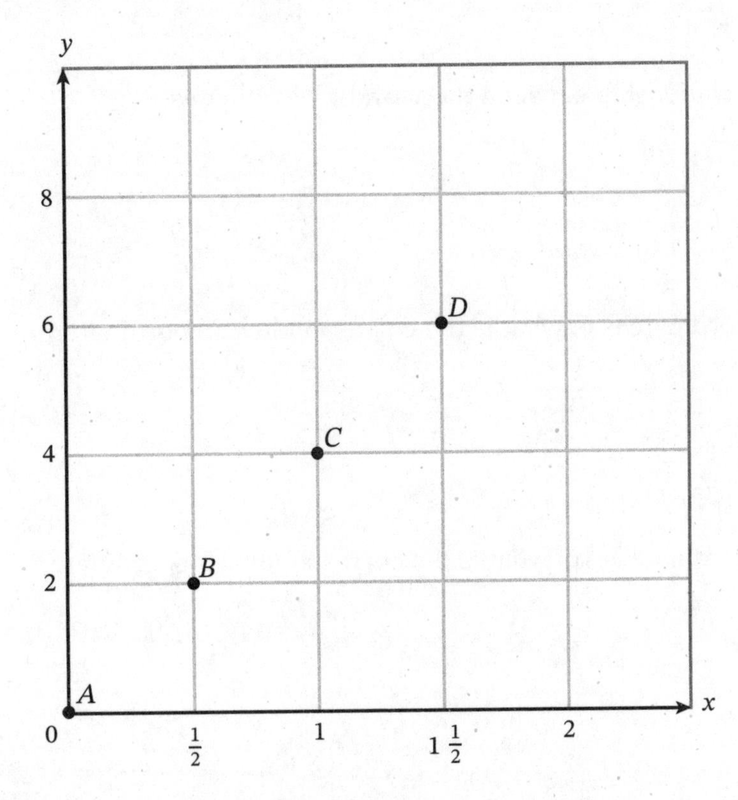

a. Write the x- and y-coordinates and ordered pairs for points A, B, C, and D.

Point	x-Coordinate	y-Coordinate	Ordered Pair

b. Every time $\frac{1}{2}$ is added to an x-coordinate, ______ is added to the corresponding y-coordinate.

c. Describe the movement to get from point C to point D.

d. Describe the relationship between the x- and y-coordinates.

e. When the x-coordinate is 6, what is the corresponding y-coordinate?

f. When the y-coordinate is 16, what is the corresponding x-coordinate?

9

Name _______________ Date _______

Use the table to complete parts (a)–(e).

x-Coordinate	*y*-Coordinate	Ordered Pair
4	1	(4, 1)
8	2	(8, 2)
12	3	(12, 3)
16	4	(16, 4)

a. Plot the four ordered pairs in the coordinate plane.

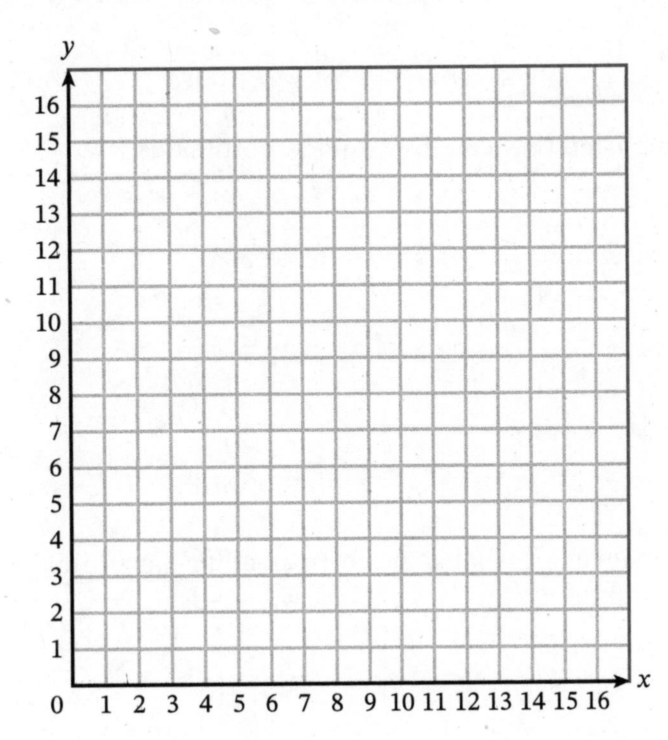

b. What is the rule for the x-coordinate?

c. What is the rule for the y-coordinate?

d. Describe the relationship between the x- and y-coordinates.

e. When the x-coordinate is 40, what is the corresponding y-coordinate?

Sprint

Write the product.

1.	8×5	
2.	$3 \times 5 \times 2$	

Number Correct: ______________

Write the product.

1.	6×1	
2.	$3 \times 2 \times 1$	
3.	3×3	
4.	$3 \times 1 \times 3$	
5.	4×4	
6.	$2 \times 2 \times 4$	
7.	6×5	
8.	$2 \times 3 \times 5$	
9.	9×5	
10.	$3 \times 3 \times 5$	
11.	5×4	
12.	$5 \times 2 \times 2$	
13.	6×6	
14.	$6 \times 2 \times 3$	
15.	7×6	
16.	$7 \times 3 \times 2$	
17.	8×8	
18.	$8 \times 2 \times 4$	
19.	9×8	
20.	$9 \times 4 \times 2$	
21.	10×9	
22.	$10 \times 3 \times 3$	

23.	4×6	
24.	6×8	
25.	$2 \times 6 \times 2$	
26.	$3 \times 8 \times 2$	
27.	8×7	
28.	9×6	
29.	$4 \times 7 \times 2$	
30.	$3 \times 6 \times 3$	
31.	9×9	
32.	2×4	
33.	$3 \times 9 \times 3$	
34.	$1 \times 2 \times 4$	
35.	8×10	
36.	9×11	
37.	6×12	
38.	$4 \times 10 \times 2$	
39.	$3 \times 11 \times 3$	
40.	$3 \times 12 \times 2$	
41.	9×12	
42.	12×12	
43.	$3 \times 12 \times 3$	
44.	$6 \times 12 \times 2$	

B

Number Correct: ______________

Improvement: ______________

Write the product.

1.	4×1	
2.	$2 \times 2 \times 1$	
3.	3×3	
4.	$3 \times 1 \times 3$	
5.	4×3	
6.	$2 \times 2 \times 3$	
7.	6×4	
8.	$2 \times 3 \times 4$	
9.	9×4	
10.	$3 \times 3 \times 4$	
11.	4×4	
12.	$4 \times 2 \times 2$	
13.	5×6	
14.	$5 \times 2 \times 3$	
15.	6×6	
16.	$6 \times 3 \times 2$	
17.	7×8	
18.	$7 \times 2 \times 4$	
19.	8×8	
20.	$8 \times 4 \times 2$	
21.	9×9	
22.	$9 \times 3 \times 3$	

23.	4×5	
24.	6×7	
25.	$2 \times 5 \times 2$	
26.	$3 \times 7 \times 2$	
27.	8×6	
28.	9×5	
29.	$4 \times 6 \times 2$	
30.	$3 \times 5 \times 3$	
31.	9×8	
32.	2×3	
33.	$3 \times 8 \times 3$	
34.	$1 \times 2 \times 3$	
35.	6×10	
36.	4×11	
37.	4×12	
38.	$3 \times 10 \times 2$	
39.	$2 \times 11 \times 2$	
40.	$2 \times 12 \times 2$	
41.	8×12	
42.	10×12	
43.	$4 \times 12 \times 2$	
44.	$5 \times 12 \times 2$	

10

Name ______________________ Date __________

1. Use one color to plot points with the coordinates shown in table A. Use another color to plot points with the coordinates shown in table B.

Table A

x-Coordinate	y-Coordinate
1	2
3	6
5	10

Table B

x-Coordinate	y-Coordinate
1	7
3	11
5	15

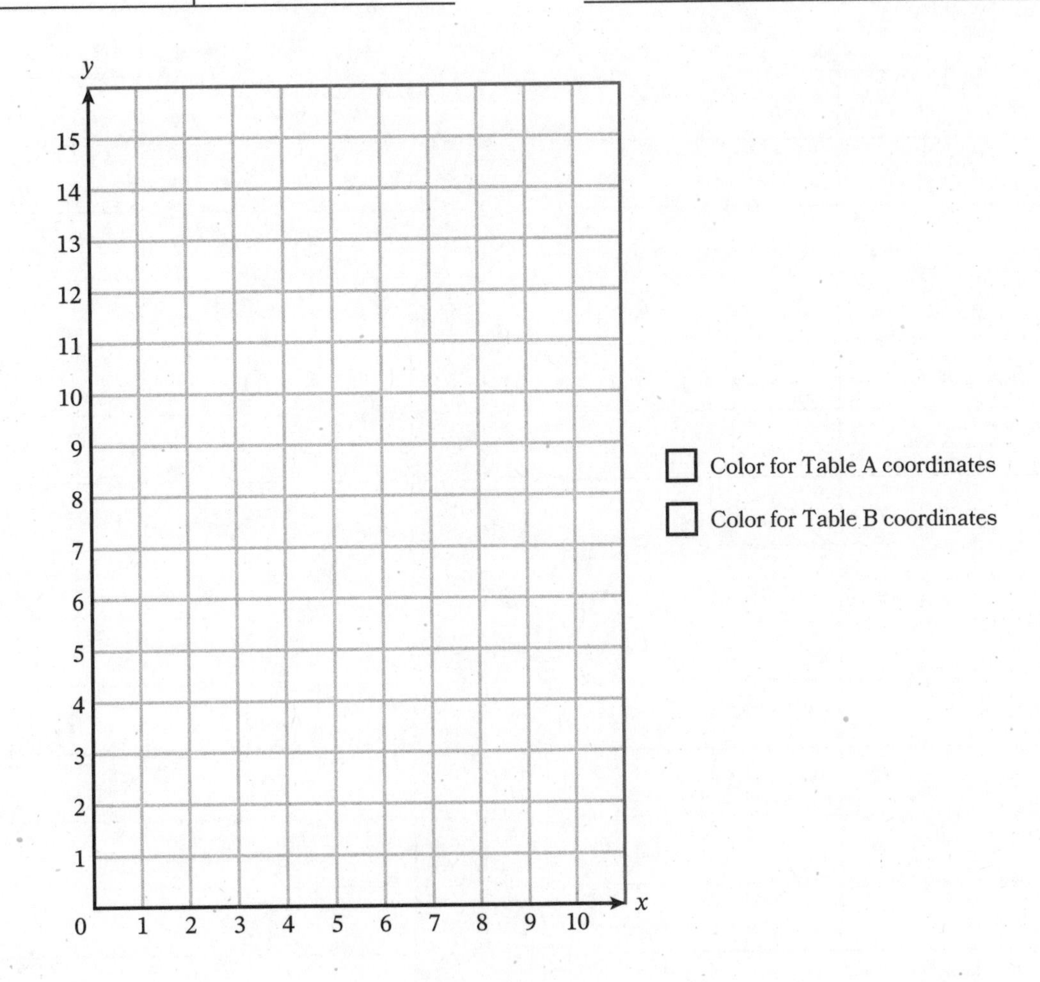

2. Complete the table.

x-Coordinate	Multiply by 3	Subtract 1	*y*-Coordinate	Ordered Pair
3	$3 \times 3 = 9$	$9 - 1 = 8$	8	(3, 8)
5				
7				
9				

Name ______________________ Date __________

1. Use the table to complete parts (a)–(c).

x-Coordinate	y-Coordinate
2	6
4	12
6	18

a. What is the rule for the x-coordinate?

b. What is the rule for the y-coordinate?

c. Fill in the blank to describe the relationship between the x- and y-coordinates.

The y-coordinate is ______ times as much as the corresponding x-coordinate.

2. Use the table to complete parts (a) and (b).

   a. Complete the table.

x-Coordinate	Multiply by 3	Subtract 2	y-Coordinate	Ordered Pair
2	$2 \times 3 = 6$	$6 - 2 = 4$	4	(2, 4)
4				
6				
8				

   b. When the x-coordinate is 9, what is the corresponding y-coordinate?

3. Use the graph to complete parts (a)–(c).

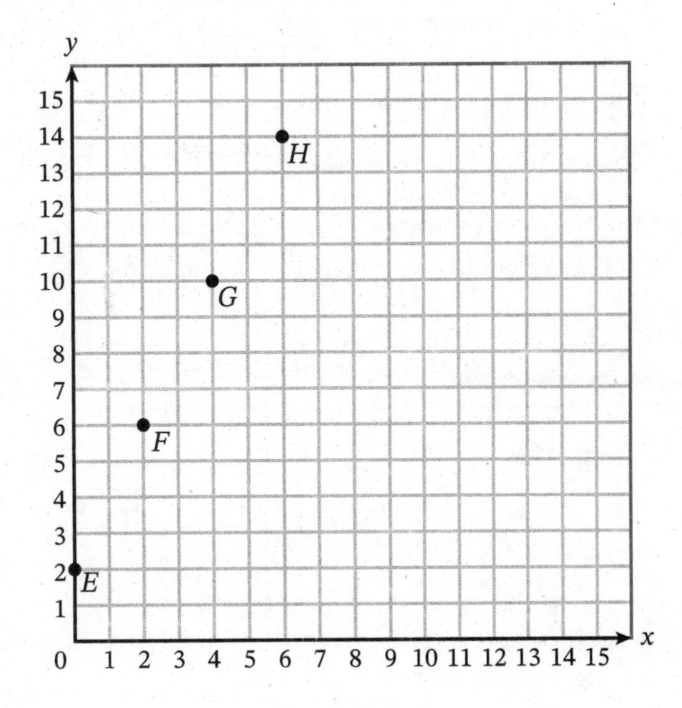

a. Complete the table.

x-Coordinate	Multiply by 2	__________	y-Coordinate
0	$0 \times 2 = 0$	$0 +$ _____ $= 2$	2
2	$2 \times 2 = 4$	$4 +$ _____ $= 6$	6
4	$4 \times 2 = 8$	$8 +$ _____ $= 10$	10
6	$6 \times 2 = 12$	$12 +$ _____ $= 14$	14

b. Describe the relationship between the x- and y-coordinates.

c. When the x-coordinate is 8, what is the corresponding y-coordinate?

4. Label each graph or table with the letter of the statement that correctly describes the relationship between the x- and y-coordinates.

A. Multiply the x-coordinates by $\frac{1}{2}$ to get the corresponding y-coordinates.

B. Multiply the x-coordinates by $\frac{1}{2}$ and then add 2 to get the corresponding y-coordinates.

C. Multiply the x-coordinates by 2 and then subtract 1 to get the corresponding y-coordinates.

D. Add 2 to the x-coordinates to get the corresponding y-coordinates.

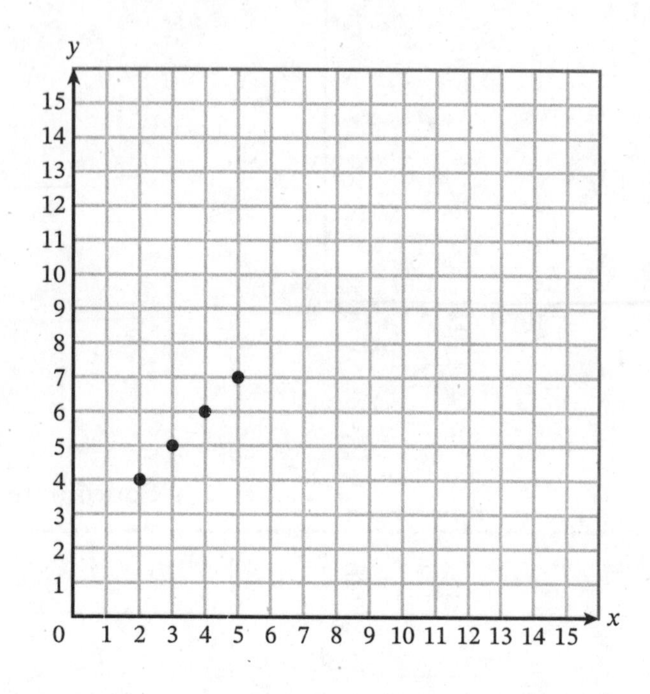

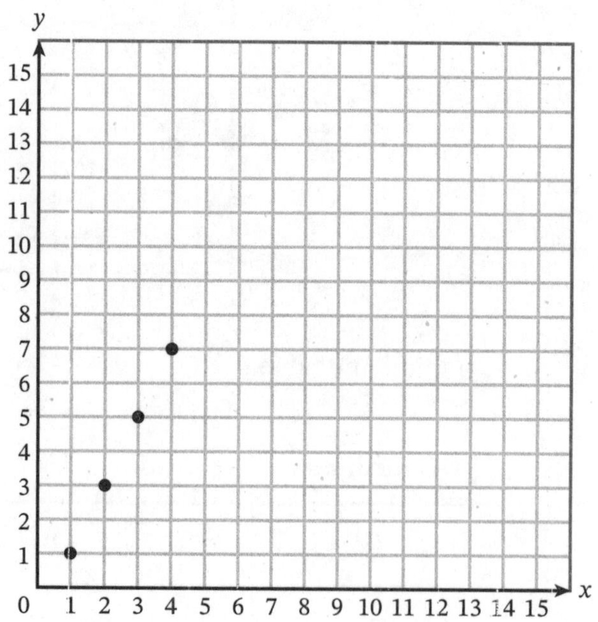

x-Coordinate	y-Coordinate	Ordered Pair
2	1	(2, 1)
6	3	(6, 3)
10	5	(10, 5)
14	7	(14, 7)

x-Coordinate	y-Coordinate	Ordered Pair
2	3	(2, 3)
6	5	(6, 5)
10	7	(10, 7)
14	9	(14, 9)

10

Name
Date

Use the graph shown to complete parts (a)–(e).

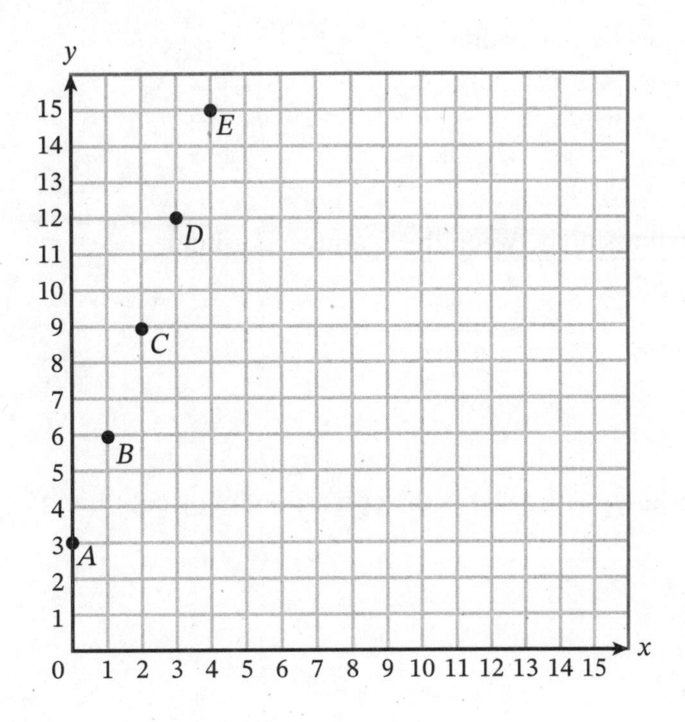

a. Complete the table.

Point	x-Coordinate	y-Coordinate	Ordered Pair
A			
B			
C			
D			
E			

b. What is the rule for the x-coordinate?

c. What is the rule for the y-coordinate?

d. Describe the relationship between the x- and y-coordinates.

e. When the x-coordinate is 5, what is the corresponding y-coordinate?

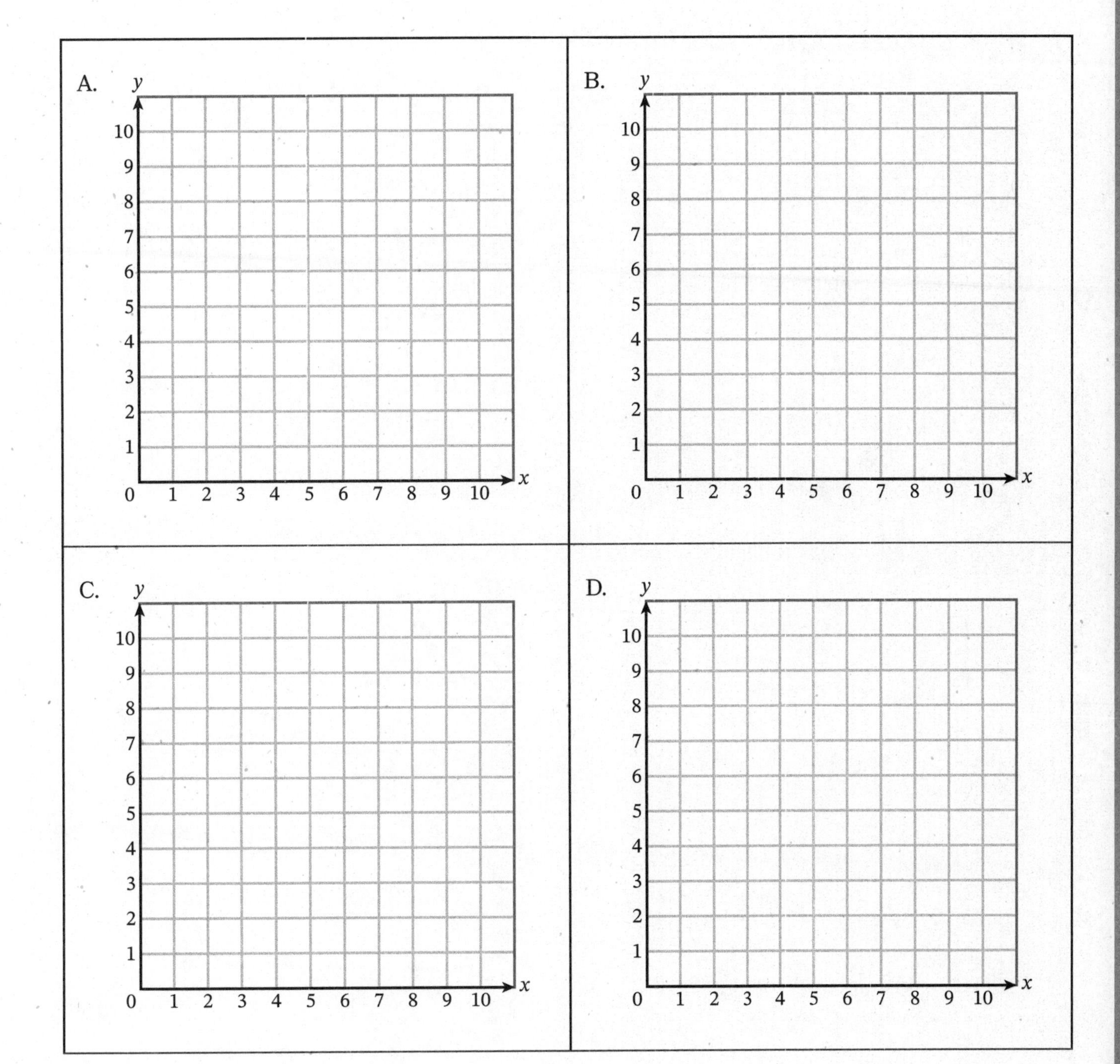
A.
y
10
9
8
7
6
5
4
3
2
1
0
1 2 3 4 5 6 7 8 9 10
x
B.
y
10
9
8
7
6
5
4
3
2
1
0
1 2 3 4 5 6 7 8 9 10
x
C.
y
10
9
8
7
6
5
4
3
2
1
0
1 2 3 4 5 6 7 8 9 10
x
D.
y
10
9
8
7
6
5
4
3
2
1
0
1 2 3 4 5 6 7 8 9 10
x

11

Name _______________ Date _______

1. Use the graphs to complete parts (a)–(d).

Use a straightedge to draw a line through the points in each coordinate plane in parts (a) and (b).

a.

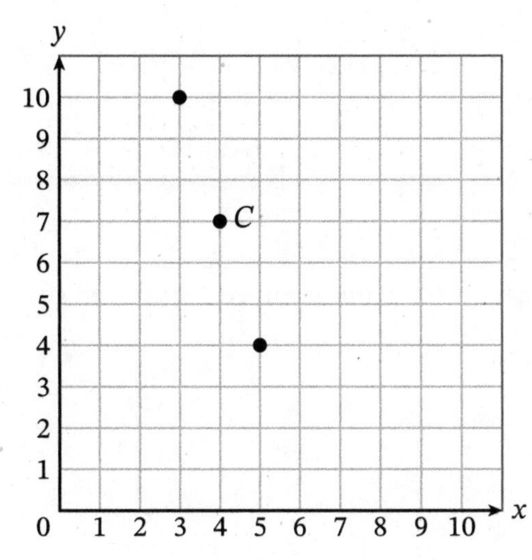

b.

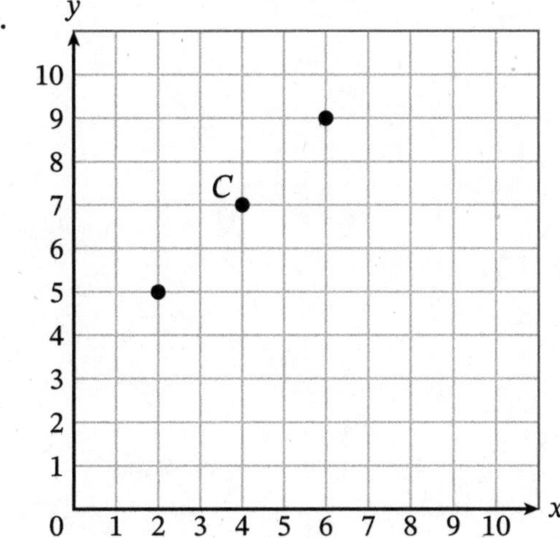

c. In both graphs, point C is located at (______, ______).

d. Draw a line through point C that is different from the lines shown in parts (a) and (b).

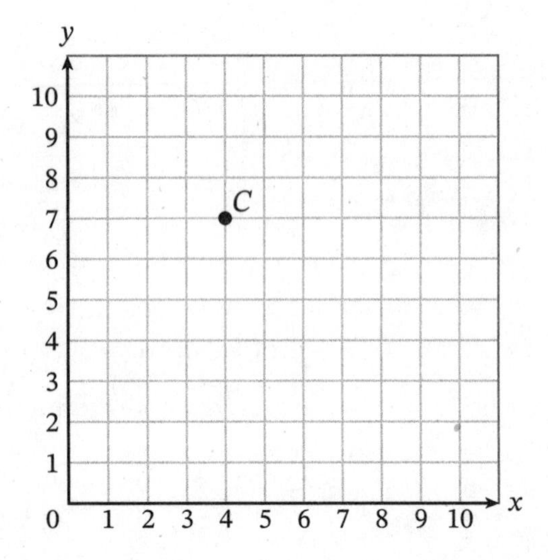

2. Use the graph of point M to complete parts (a)–(d).

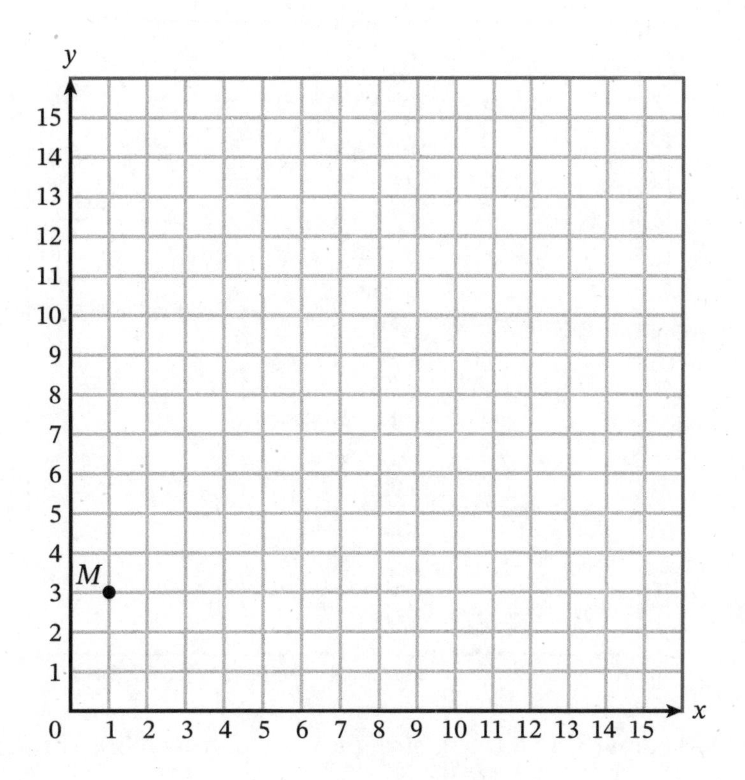

a. Point M is located at (1, 3). One possible relationship between the x- and y-coordinates of this point is that the y-coordinate is 2 more than the corresponding x-coordinate.

Write three more ordered pairs with this relationship between the x- and y-coordinates.

b. Plot the points from part (a). Use a straightedge to draw a line through the points.

c. Consider another line that passes through point M. Write a relationship between the x- and y-coordinates of points that are on the new line.

d. Write three more ordered pairs with the relationship you wrote in part (c). Plot these points. Use a straightedge to draw a line through the points.

3. Use the coordinate plane to complete parts (a)–(d).

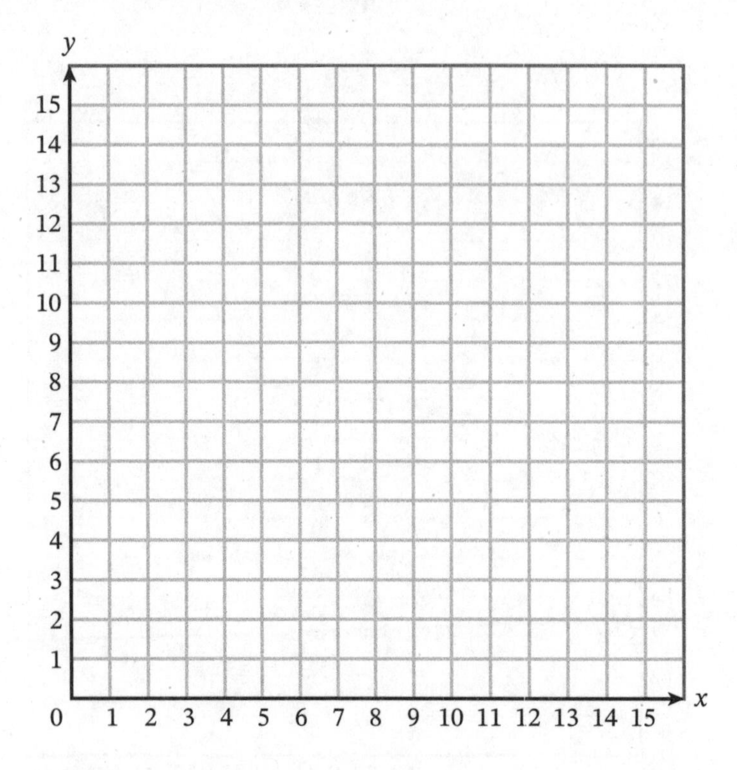

a. Plot points (1, 3) and (5, 15). Use a straightedge to draw a line through the points.

b. Each y-coordinate is 3 times as much as the corresponding x-coordinate for all points that lie on the line. Name two other points on the line.

c. Does the point $\left(\frac{2}{3}, 2\right)$ lie on the line? How do you know?

d. Sort the following ordered pairs by writing them in the correct column of the table.

$(3, 12)$ $(5, 15)$ $(9, 27)$ $(3, 0)$ $(9, 3)$ $\left(1\frac{1}{3}, 4\right)$ $(1, 3)$ $(0, 3)$ $(7, 21)$

Points on the Line	Points Not on the Line

11

Name _______________ Date _______________

Use the coordinate plane to complete parts (a)–(c).

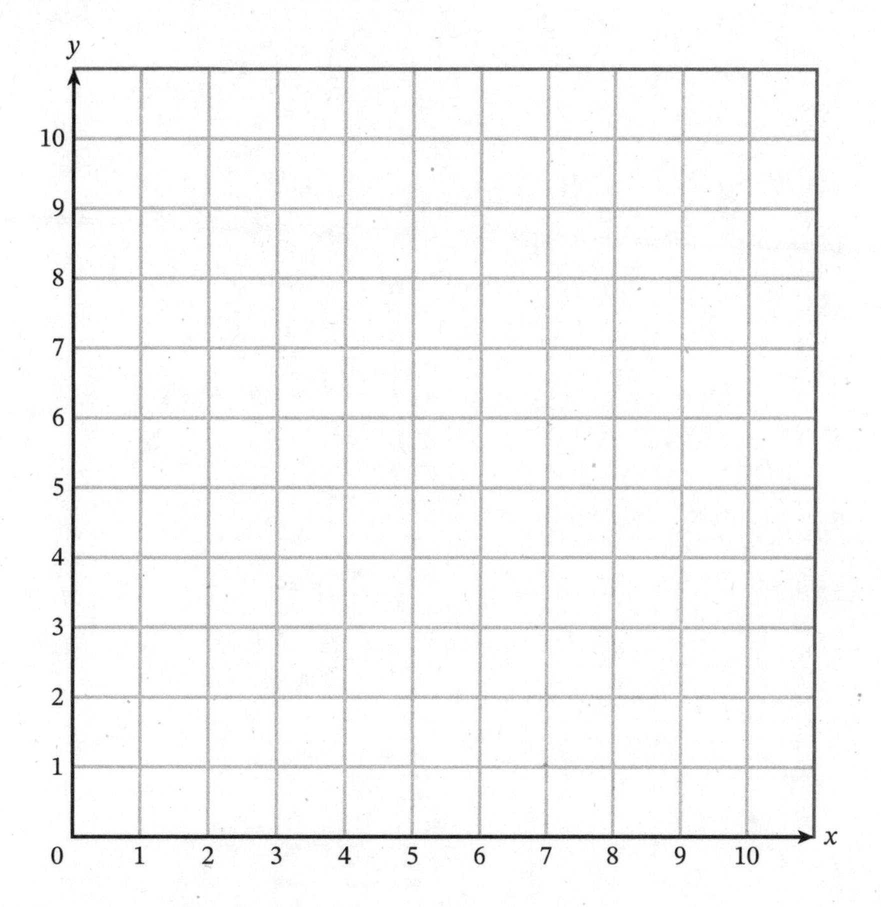

a. Plot the points (3, 1) and (8, 6) in the coordinate plane.

b. Use a straightedge to draw a line through the points.

c. Write ordered pairs for three other points that lie on the line.

12

Name ____________________ Date ____________

1. Use the graph to complete parts (a) and (b).

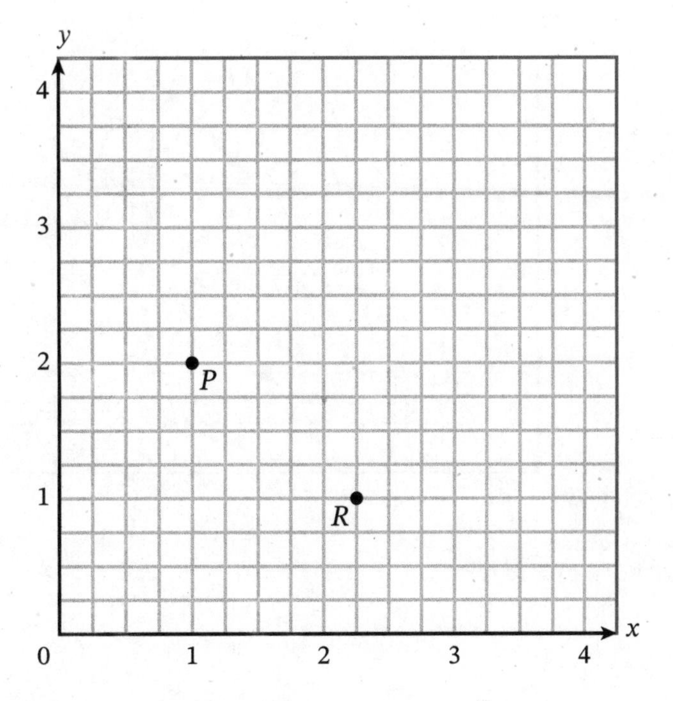

a. Plot point Q so that $\overline{PQ}$ is vertical and has a length of $1\frac{3}{4}$ units. Use a straightedge to draw $\overline{PQ}$. What is the ordered pair for point Q?

b. Plot point S so that $\overline{RS}$ is horizontal and has a length of $\frac{1}{2}$ unit. Use a straightedge to draw $\overline{RS}$. What is the ordered pair for point S?

2. Use the graph to complete parts (a)–(d).

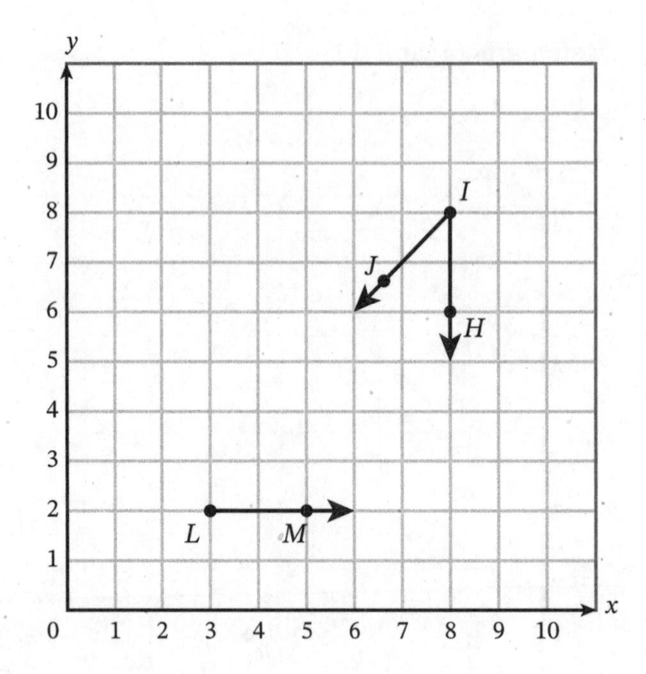

a. What is the measure of $\angle HIJ$?

b. Plot point N so that $\angle MLN$ is a right angle. Draw $\overrightarrow{LN}$. What is the measure of $\angle MLN$?

c. Plot point O so that $\angle MLO$ is an obtuse angle. Draw $\overrightarrow{LO}$. What is the measure of $\angle MLO$?

d. Plot point P so that $\angle MLP$ is an acute angle. Draw $\overrightarrow{LP}$. What is the measure of $\angle MLP$?

Name ______________________ Date __________

1. Use the coordinate plane to complete parts (a)–(c).

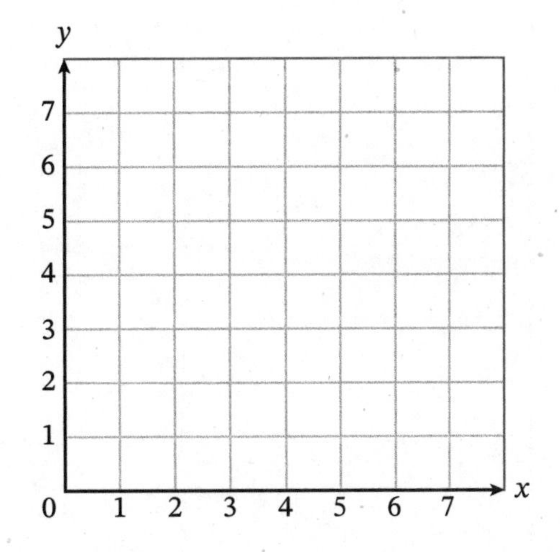

a. Draw a horizontal line segment with a length of 4 units and one endpoint at (2, 4).

b. Write the ordered pair for the other endpoint.

c. All points on the line segment have the same ______-coordinate but different ______-coordinates.

2. Use the graph of $\overrightarrow{RC}$ to complete parts (a)–(d).

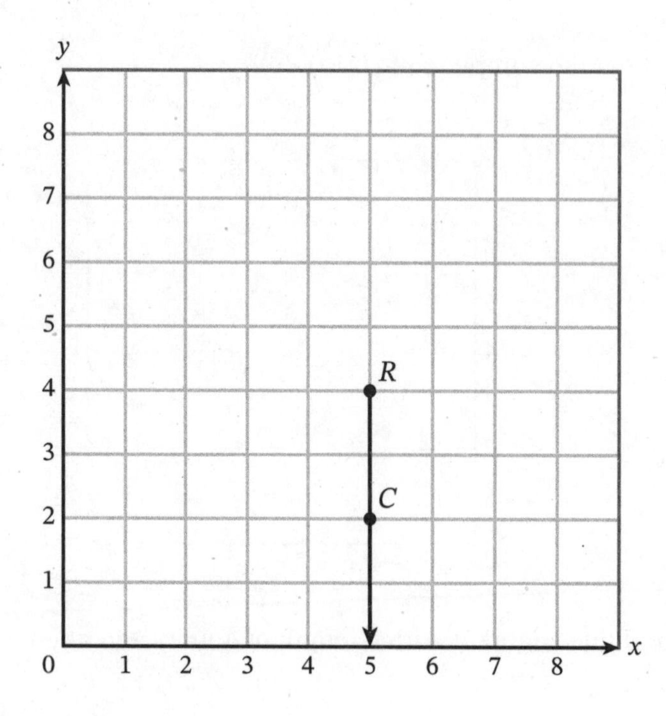

a. Plot point G so that $\angle CRG$ is a right angle.

b. Plot point F so that $\angle CRF$ is an obtuse angle.

c. Plot point M so that $\angle CRM$ is an acute angle.

d. Explain how you know $\angle CRM$ is an acute angle.

3. Use the coordinate plane to complete parts (a)–(e).

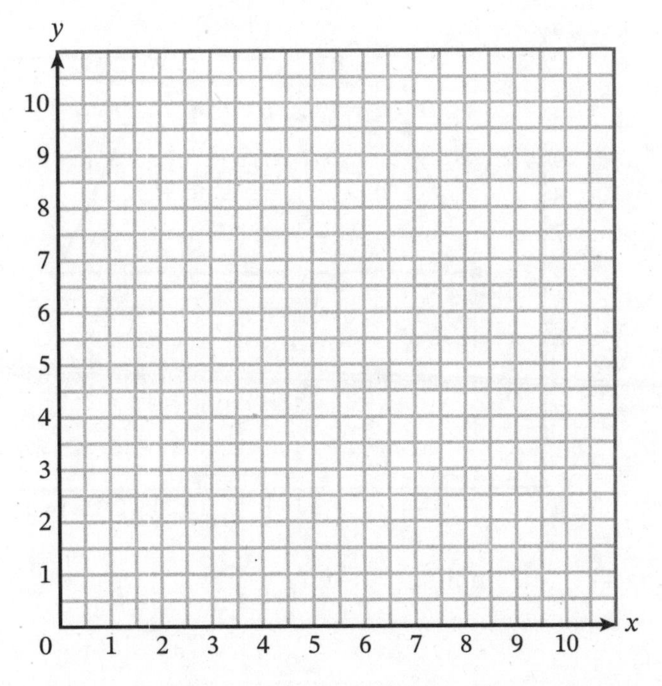

a. Plot and label the following vertices: $D\left(1\frac{1}{2}, 2\right)$, $E\left(1\frac{1}{2}, 8\right)$, $F(4, 8)$, $G(4, 2)$. Connect the vertices to create polygon $DEFG$.

b. Which side has the same length as $\overline{DE}$? Explain.

c. Which sides are perpendicular to $\overline{DG}$?

d. Describe the angles of this polygon.

e. What is the most specific name for polygon $DEFG$?

4. Use the graph of polygon *FGHI* to complete parts (a)–(e).

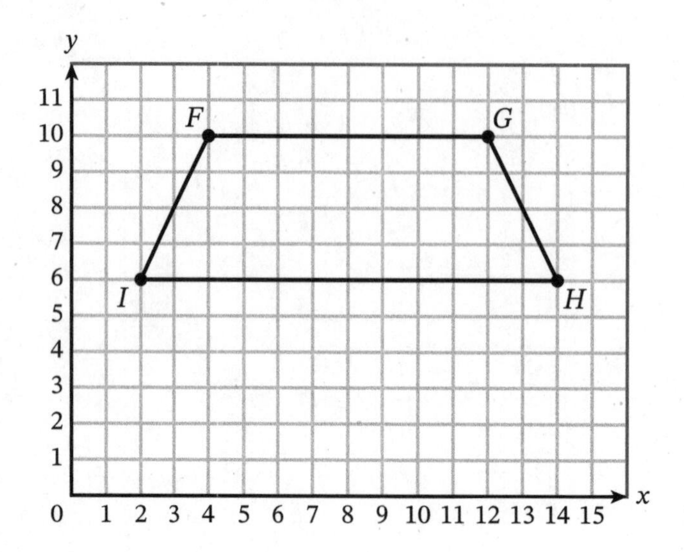

a. The length of $\overline{FG}$ is ______ units.

b. The length of $\overline{IH}$ is ______ units.

c. $\overline{FG}$ is parallel to ______.

d. Describe the angles of this polygon.

e. What is the most specific name for polygon *FGHI*?

5. Use the coordinate plane to complete parts (a)–(e).

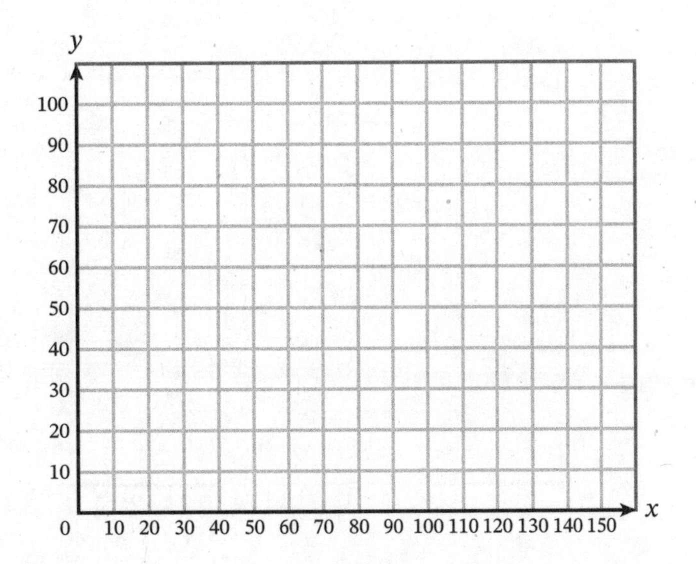

a. Plot and label the following vertices: $L(110, 30)$, $M(130, 90)$, $N(50, 90)$, $O(30, 30)$. Connect the vertices to create polygon $LMNO$.

b. What is the length of $\overline{NM}$?

c. Is the length of $\overline{OL}$ equal to the length of $\overline{NM}$? Explain.

d. Describe the angles of this polygon.

e. What is the most specific name for polygon $LMNO$?

12

Name _______________ Date _______

Use the coordinate plane for parts (a)–(g).

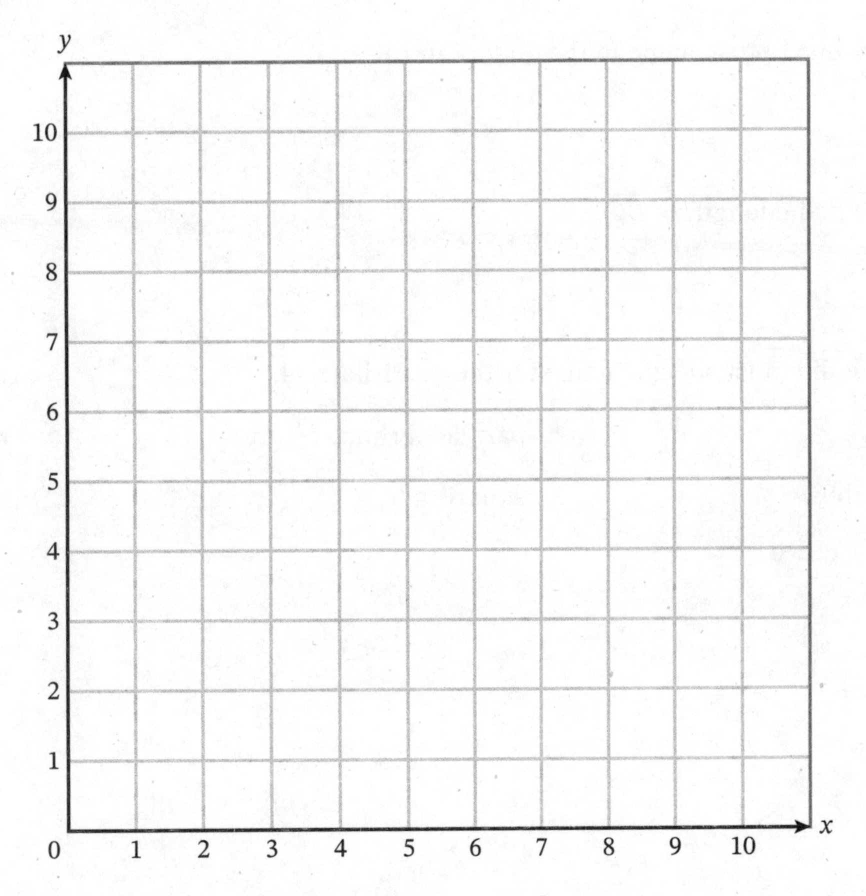

a. Plot the given points in the coordinate plane.

$A(2, 2)$

$B(2, 9)$

$C(8, 9)$

$D(5, 2)$

b. Connect the points to make a quadrilateral.

c. Write one acute angle in the quadrilateral.

d. Write one right angle in the quadrilateral.

e. Write one obtuse angle in the quadrilateral.

f. What is the length of $\overline{BC}$?

g. Circle the most specific name for the quadrilateral.

kite	parallelogram	rectangle
rhombus	square	trapezoid

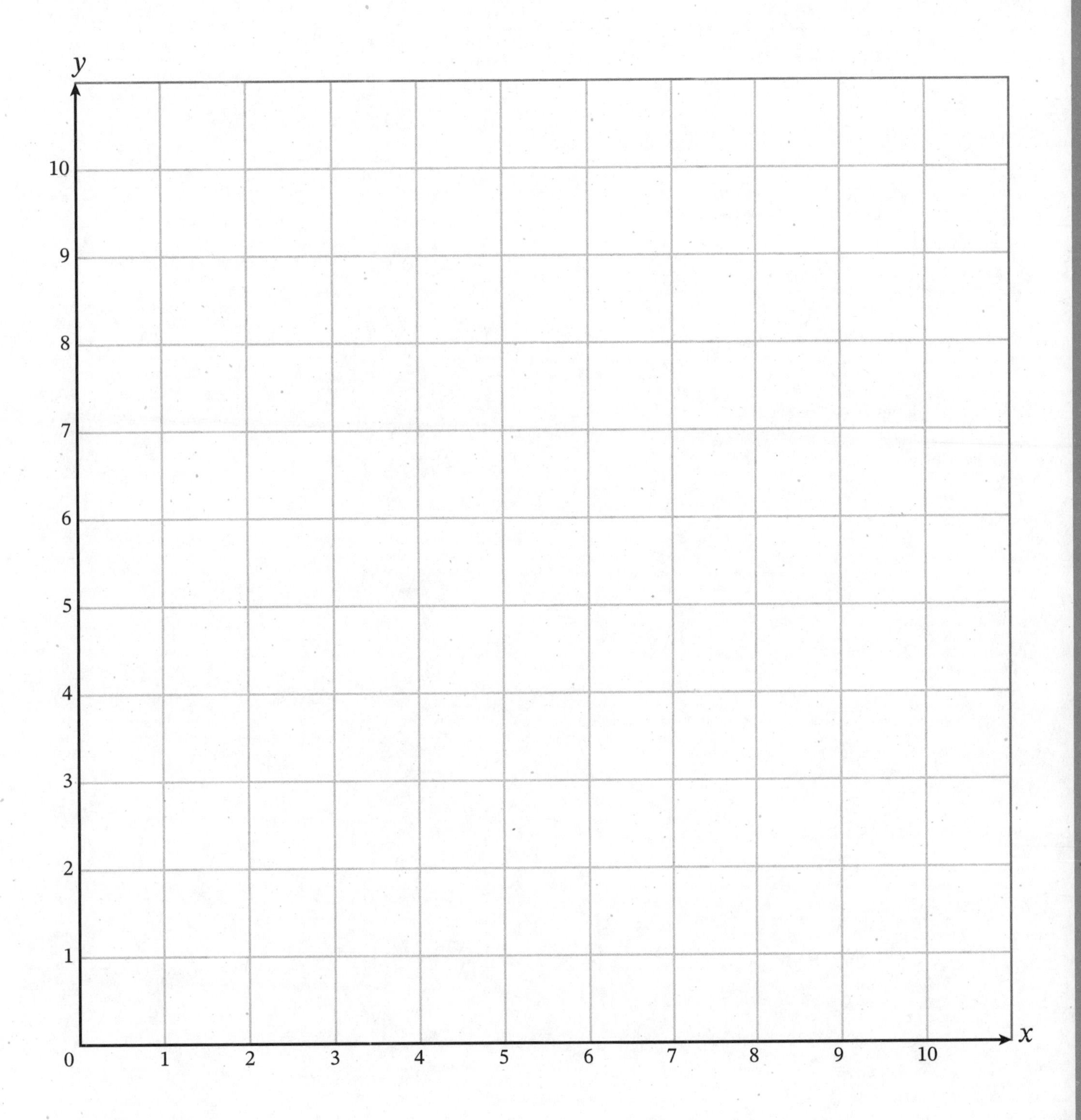
y
10
9
8
7
6
5
4
3
2
1
0
1
2
3
4
5
6
7
8
9
10
x

13

Name

Date

Consider the figure shown in the coordinate plane.

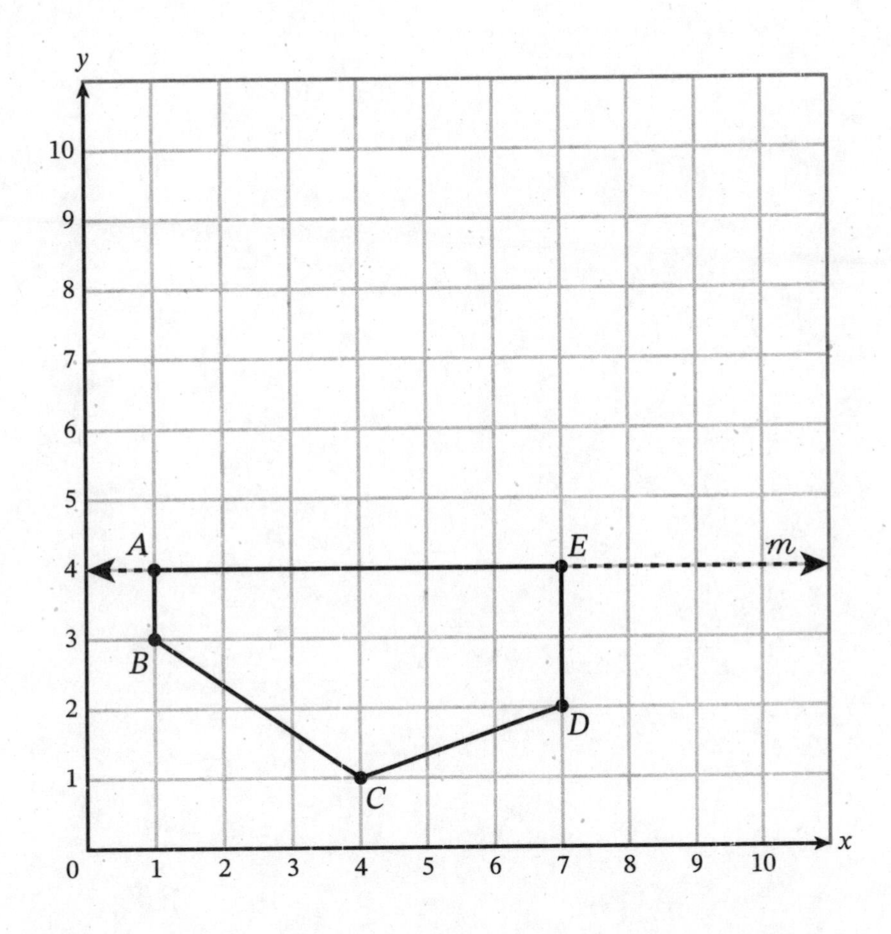

a. Create a figure that is symmetric across line m by using the given points.

b. Describe how the x- and y-coordinates for points B, C, and D relate to the x- and y-coordinates for the points that are symmetric to points B, C, and D.

Part A: Plot any two points in the coordinate plane that do not lie on the same horizontal or vertical line.

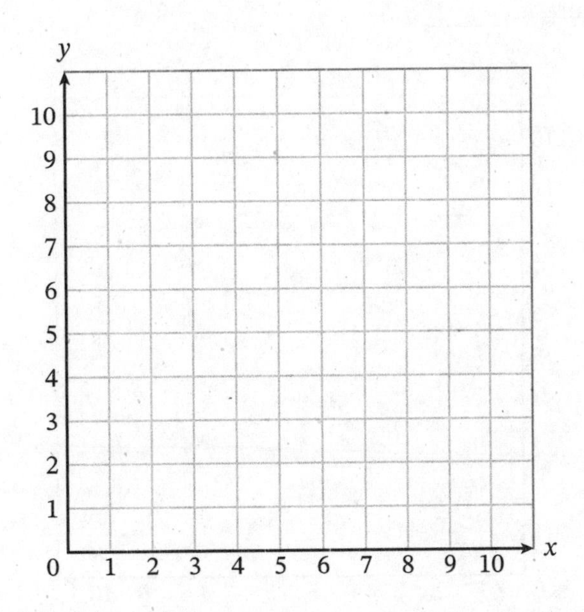

Part B: Plot any two points in the coordinate plane that lie on the same vertical line.

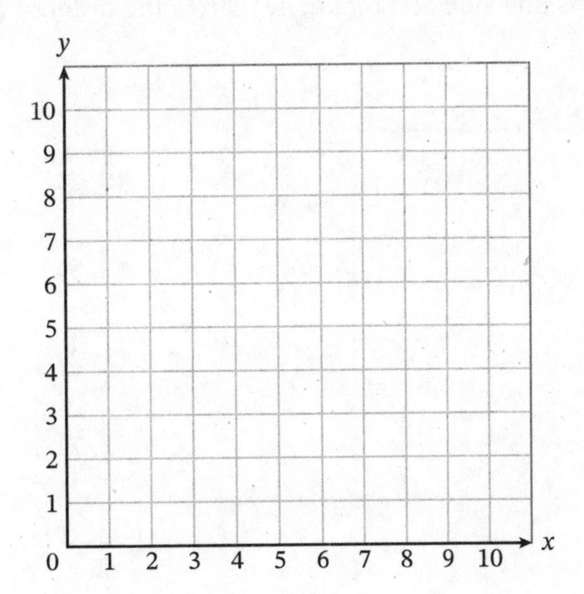

Part C: Plot any two points in the coordinate plane that lie on the same horizontal line.

Write a number between 1 and 5: ______.

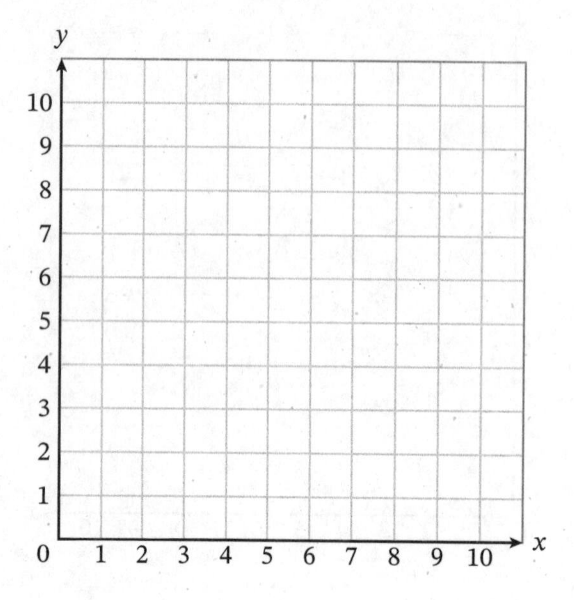

Part D: The segment shown is one side of a rectangle. Circle the ordered pair for any point that could be a vertex of the rectangle.

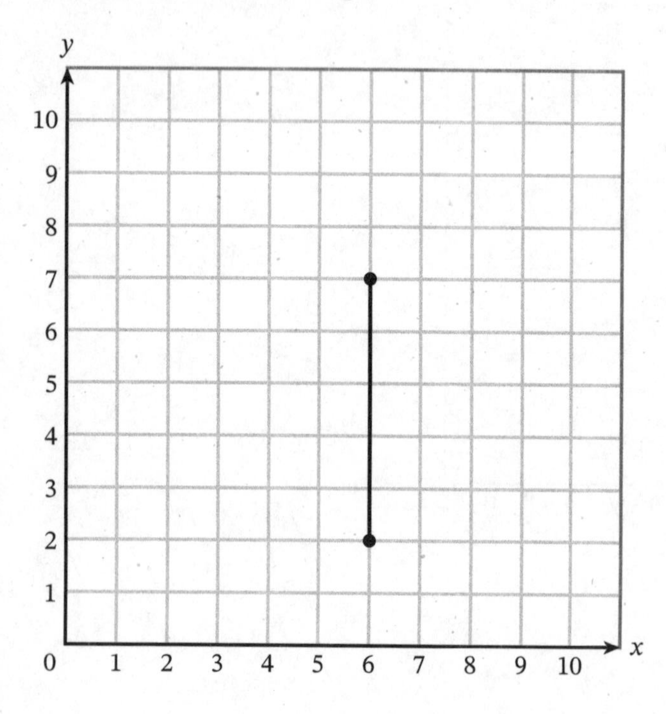

(8, 2) (8, 7)

(2, 8) (15, 2)

(1, 2) (0, 7)

(1, 7) (2, 0)

$\left(3\frac{1}{2}, 2\right)$ (12.1, 7)

Part E: A rectangle with a length of 5 units and a width of 3 units has one vertex at (4, 4) as shown.

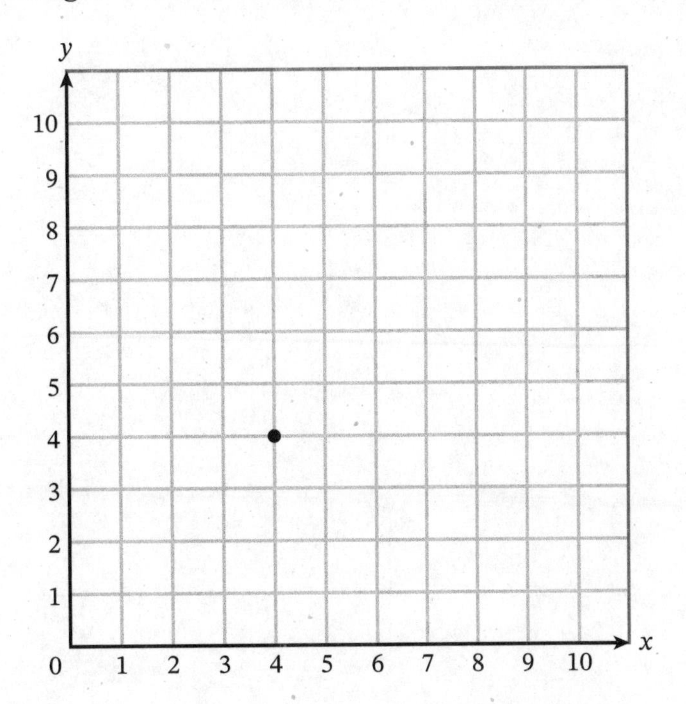

14

Name ______________________ Date ____________

1. Three of the vertices of a rectangle are $A(2, 3)$, $B(2, 8)$, and $C(6, 8)$.

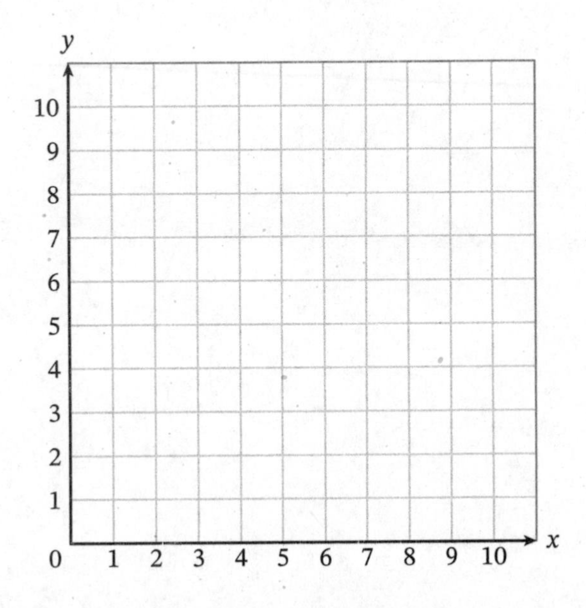

a. Plot and label the three vertices in the coordinate plane.

b. Determine the ordered pair for point D, the fourth vertex.

c. Draw rectangle $ABCD$.

d. Identify the ordered pair for a point on $\overline{CD}$ other than point C or point D.

14

Name ______________________ Date __________

1. Rectangle $MNOP$ is shown in the coordinate plane.

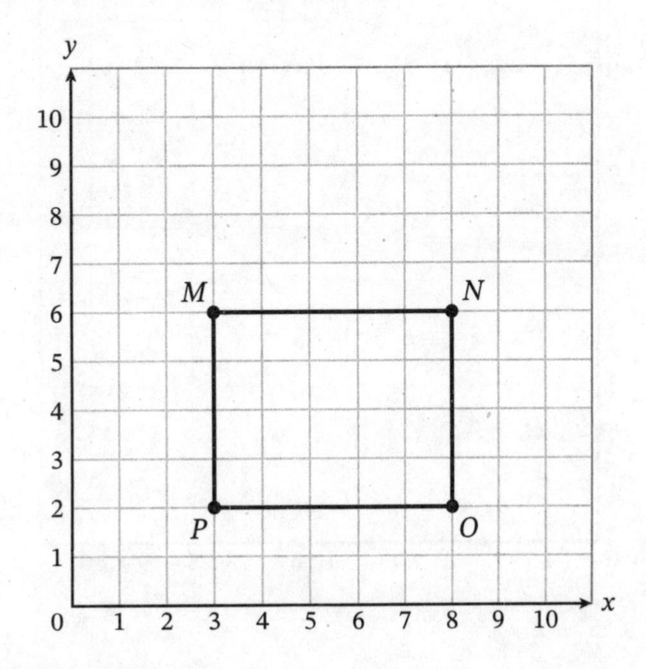

a. Circle the ordered pairs for vertices of rectangle $MNOP$.

(5, 6) (8, 5) (8, 2) (8, 6)

(3, 2) (4, 2) (3, 4) (3, 6)

b. Points M and N have the same ______-coordinate because they are on the same horizontal line.

c. Points N and O have the same ______-coordinate because they are on the same vertical line.

2. Points R, S, and T are three of the vertices of a rectangle. Plot the fourth vertex of the rectangle. Label the point U and write its ordered pair next to the point.

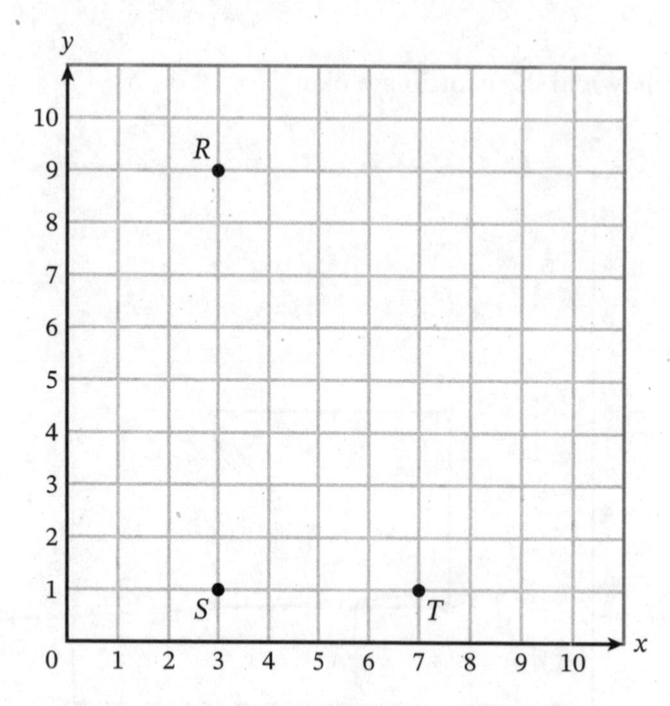

3. Points A and C are opposite vertices of a rectangle.

   a. Plot the other two vertices of the rectangle. Label the points B and D.

   b. Draw rectangle $ABCD$.

   c. What are the coordinates of points B and D?

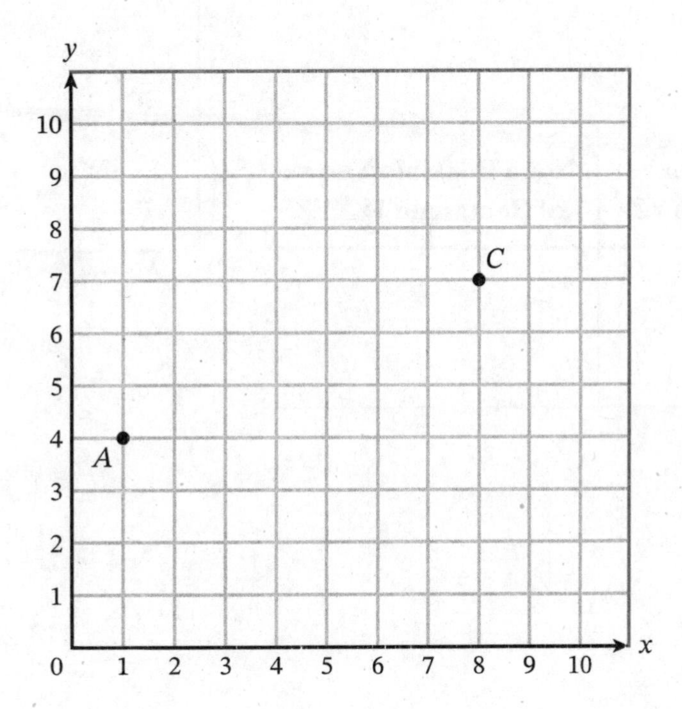

4. $\overline{WX}$ of rectangle *WXYZ* is shown in the coordinate plane. The width of rectangle *WXYZ* is 2 units.

   Determine whether each ordered pair could be the location of a vertex of rectangle *WXYZ*. Write each ordered pair in the correct column of the table.

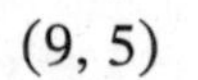

(9, 5) (9, 6) (2, 8) (9, 9)

(2, 5) (9, 8) (2, 6) (2, 9)

Possible Vertex of Rectangle *WXYZ*	Not a Possible Vertex of Rectangle *WXYZ*

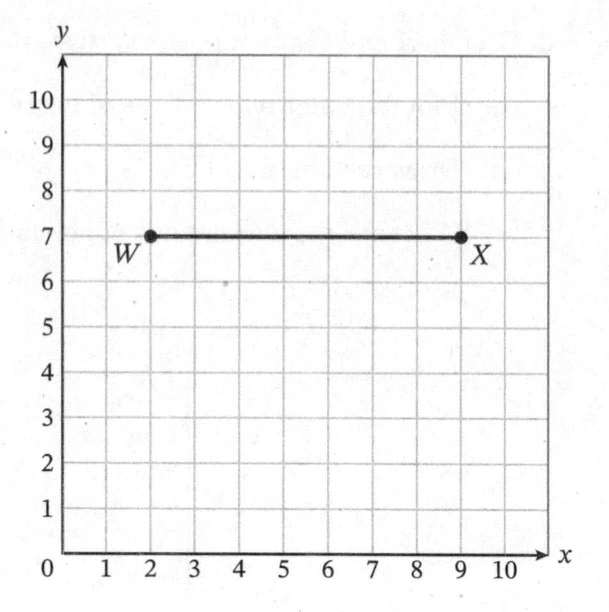

5. Point H is plotted at $(4, 5)$.

   a. Draw a rectangle with a length of 5 units and a width of 4 units. Use point H as one of the rectangle's vertices.

   b. What are the coordinates of the three other vertices of your rectangle?

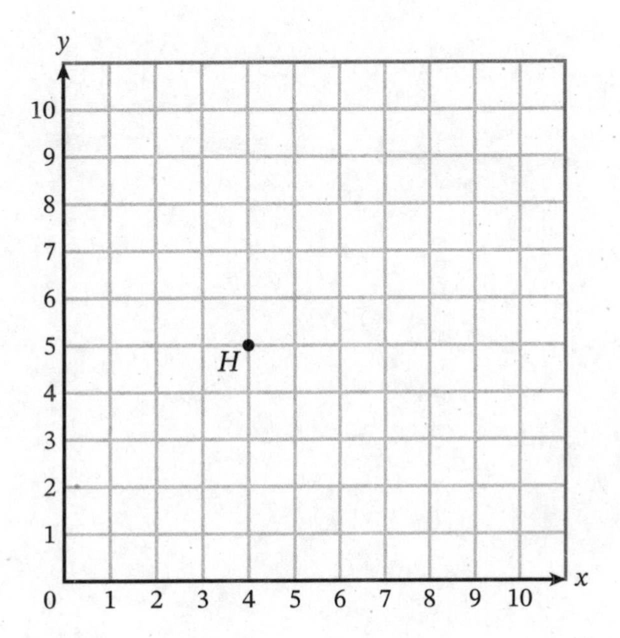

14

Name

Date

The points $A(2, 4)$ and $C(6, 7)$ are two opposite vertices of a rectangle.

a. Plot the four vertices of rectangle $ABCD$ in the coordinate plane and draw the rectangle.

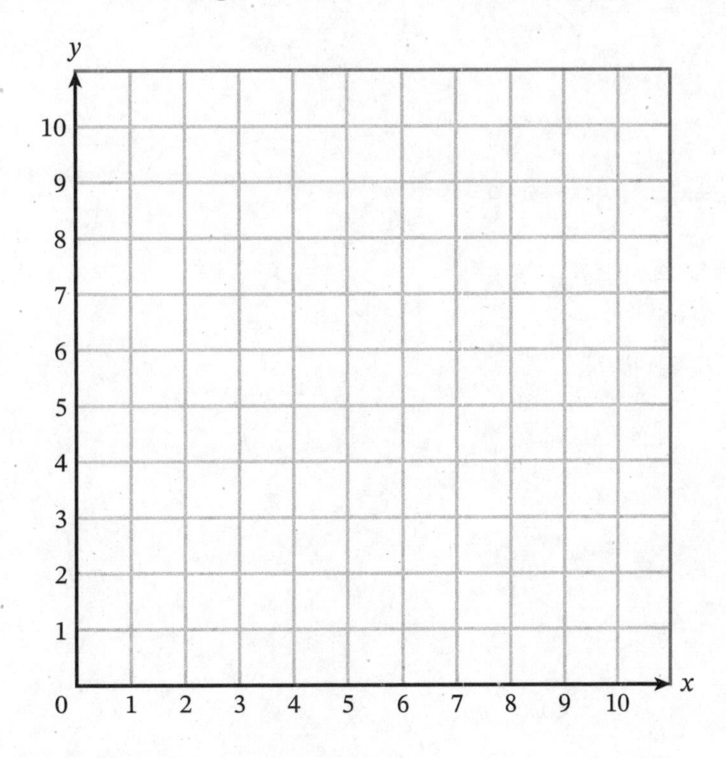

b. Write the coordinates of point B and point D.

c. What are the length and width of rectangle $ABCD$?

15

Name ______________________ Date ____________

1. Determine the perimeter of rectangle *ABCD*.

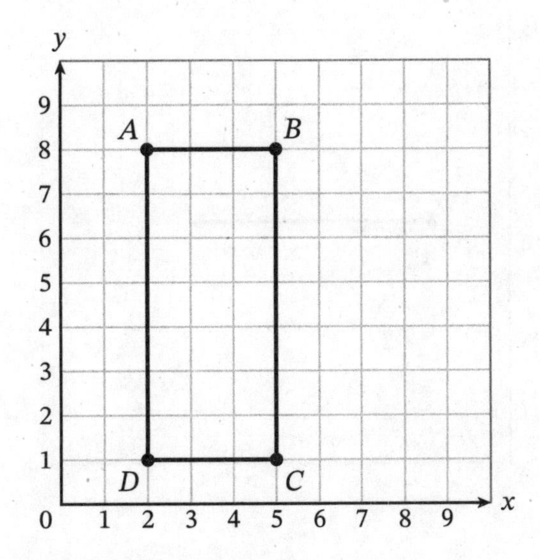

2. Determine the perimeter of rectangle *EFGH*.

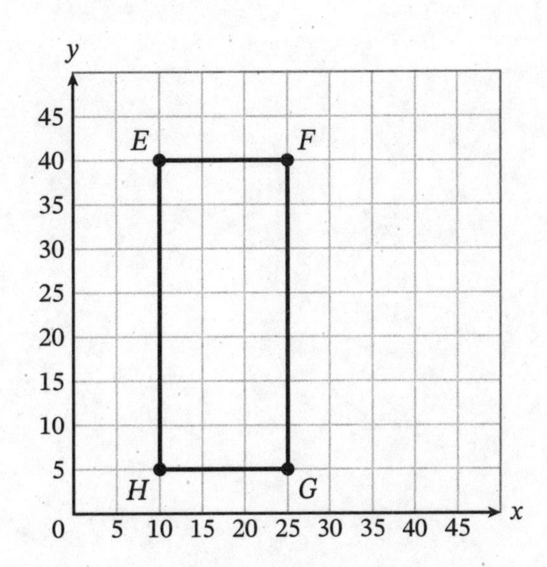

3. The graph shows one side of rectangle *MNPQ*.

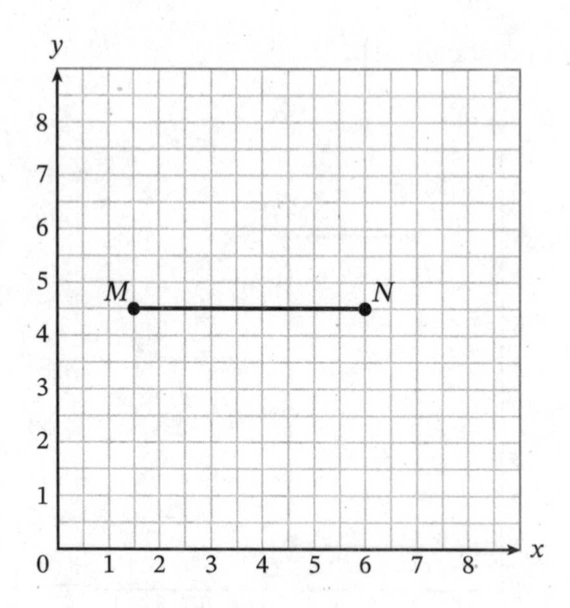

a. Rectangle *MNPQ* has a perimeter of 16 units. Plot points *P* and *Q*.

b. What are the ordered pairs for points *P* and *Q*?

4. Circle every rectangle that has an area of 36 square units.

A.

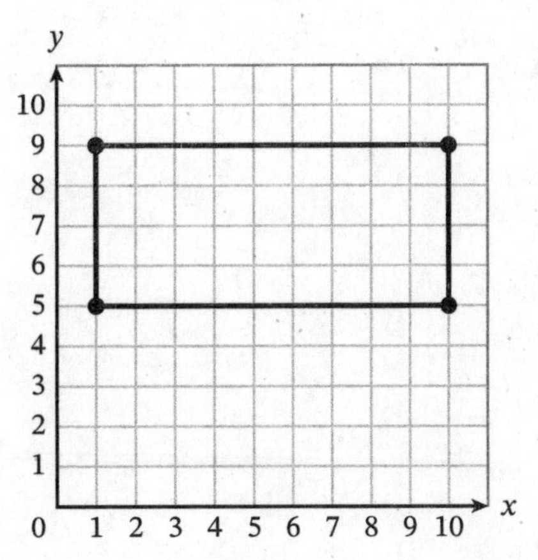

B.

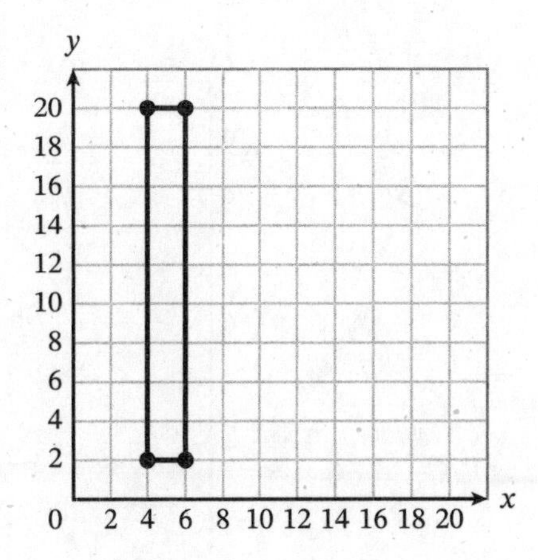

C.

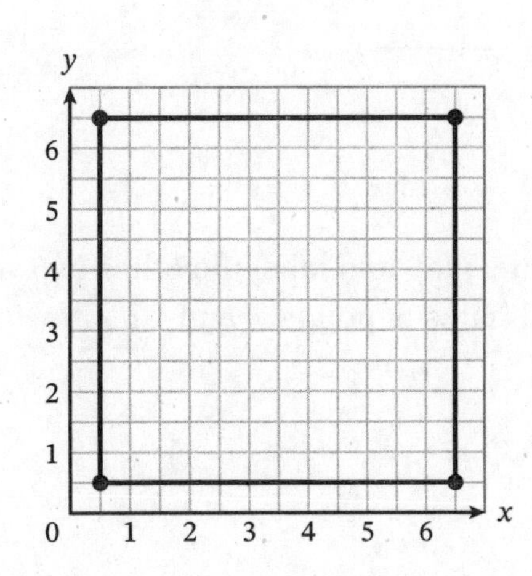

D.

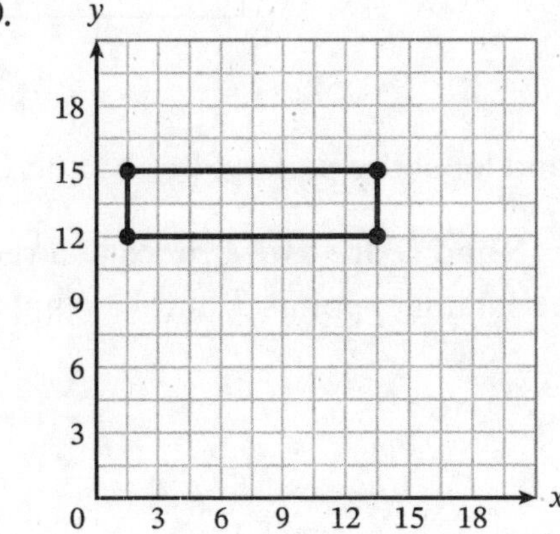

E.

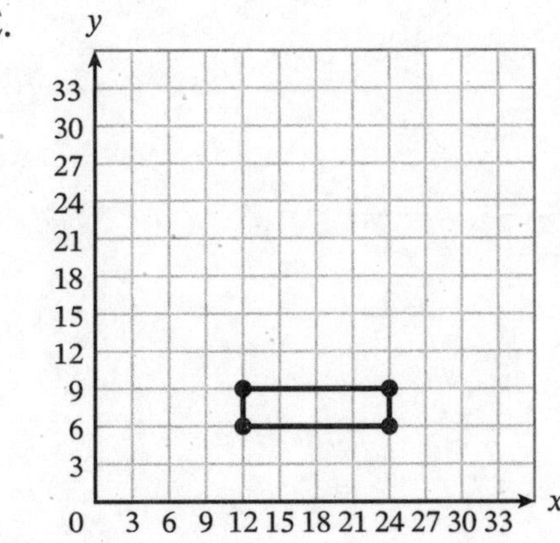

F.

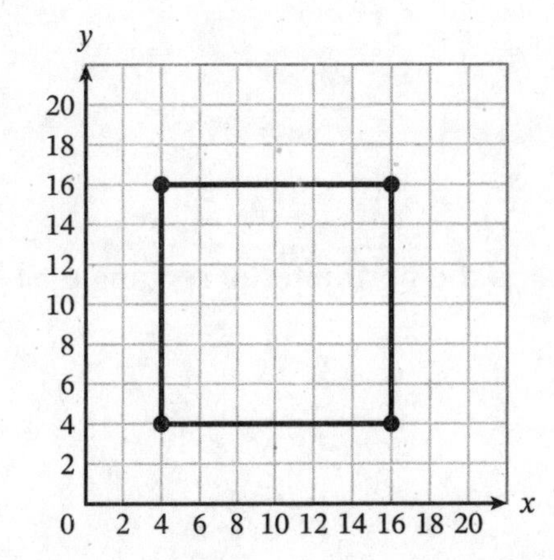

5. Use the coordinate plane to answer parts (a)–(e).

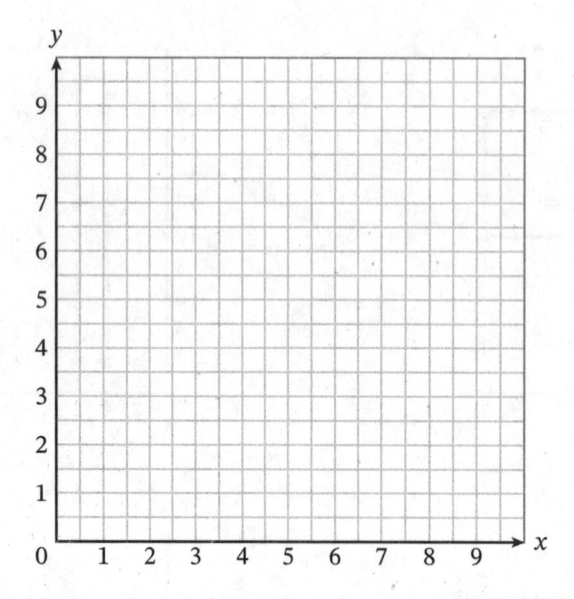

a. Plot and label the points $S\left(\frac{1}{2}, 3\right)$ and $U\left(7, 8\frac{1}{2}\right)$.

b. Points S and U are two vertices of a rectangle. Locate, plot, and label the other two vertices of the rectangle, points T and V. What are the coordinates of points T and V?

c. What are the length and width of rectangle $STUV$?

d. What is the perimeter of rectangle $STUV$?

e. What is the area of rectangle *STUV*?

15

Name ________________________ Date ____________

1. The graph shows rectangle *MNOP*.

   a. The length of $\overline{MP}$ is ______ units.

   b. The length of $\overline{PO}$ is ______ units.

   c. The length of $\overline{ON}$ is ______ units.

   d. The length of $\overline{NM}$ is ______ units.

   e. The perimeter of rectangle *MNOP* is ______ units.

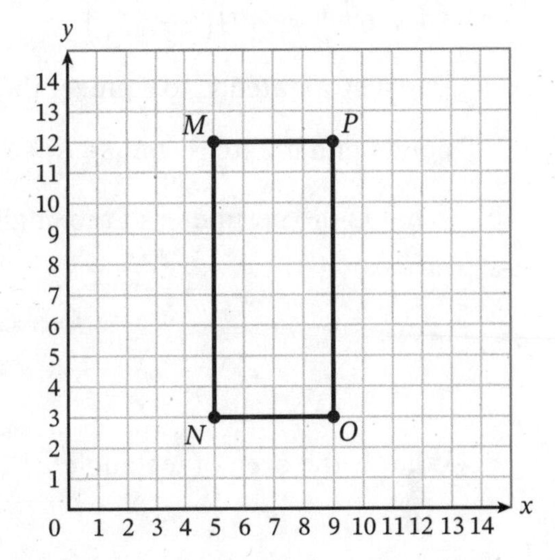

2. Rectangle *EFGH* and rectangle *HIJK* are each graphed in one of the coordinate planes shown.

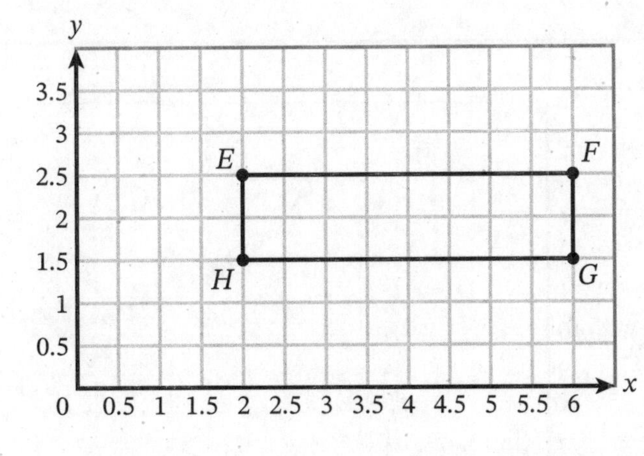

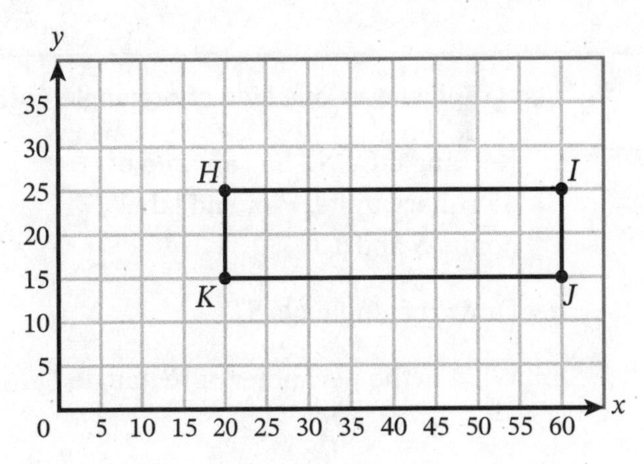

   a. The interval length of the axes of the coordinate plane with rectangle *HIJK* is ______ times as much as the interval length of the axes of the coordinate plane with rectangle *EFGH*.

   b. Which rectangle has a greater perimeter?

3. Use the coordinate plane to complete parts (a)–(c).

   a. Plot points $E\left(2\frac{1}{2}, 2\right)$, $F(8, 2)$, $G\left(8, 6\frac{1}{2}\right)$, and $H\left(2\frac{1}{2}, 6\frac{1}{2}\right)$.

   Use a straightedge to connect the points and create rectangle $EFGH$.

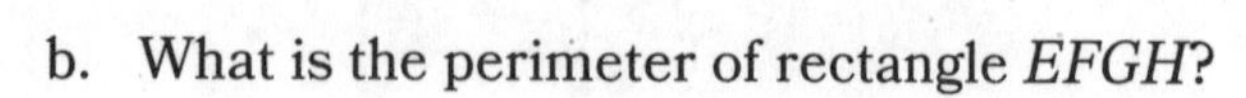

   b. What is the perimeter of rectangle $EFGH$?

   c. What is the area of rectangle $EFGH$?

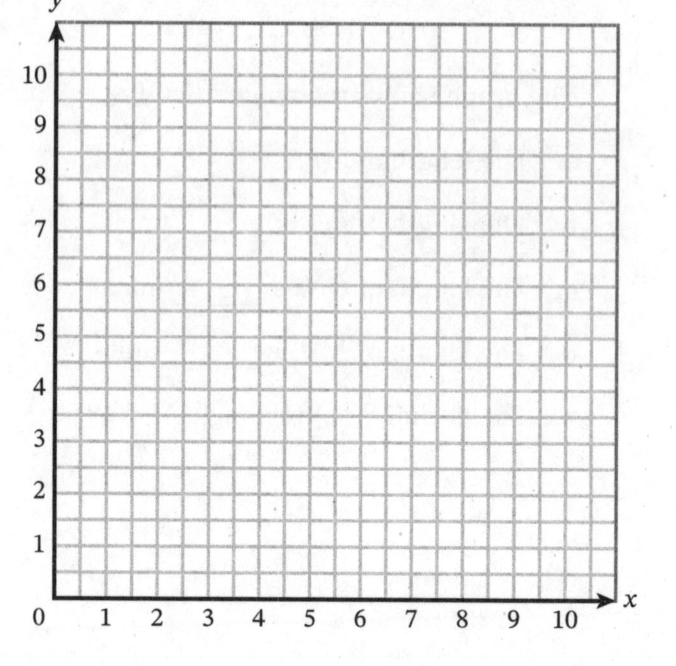

4. The graph shows one side of rectangle $QRST$.

   a. Rectangle $QRST$ has an area of 45 square units. Plot and label points S and T.

   b. Draw rectangle $QRST$.

   c. What is the perimeter of rectangle $QRST$?

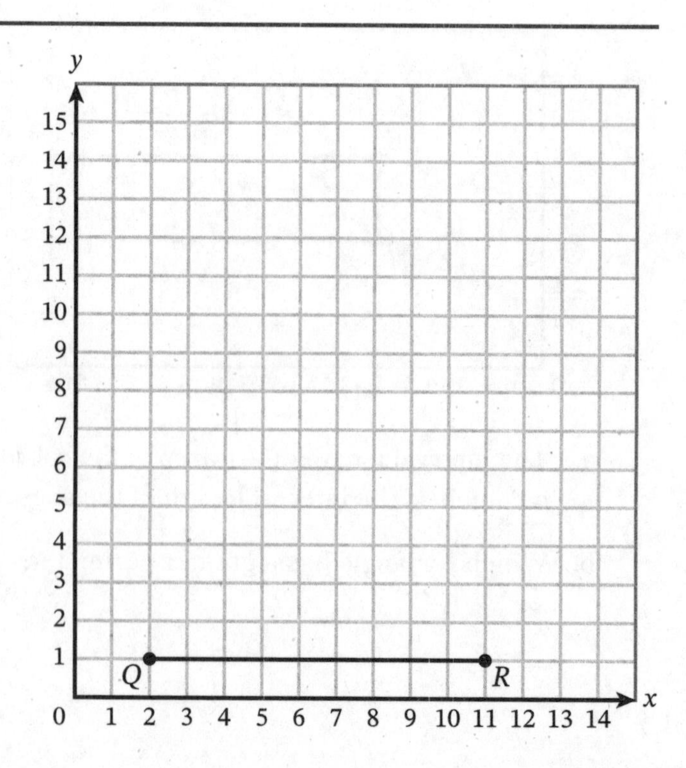

5. Tara says the perimeter of rectangle $WXYZ$ is 38 units. Adesh says the perimeter of rectangle $WXYZ$ is 19 units. Who is correct? Explain.

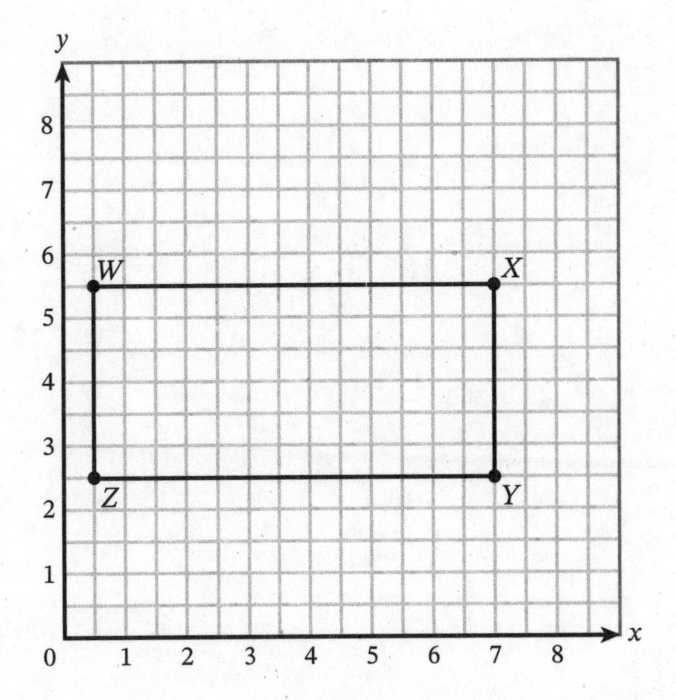

15

Name ______________________ Date __________

1. Rectangle $ABCD$ is graphed in the coordinate plane.

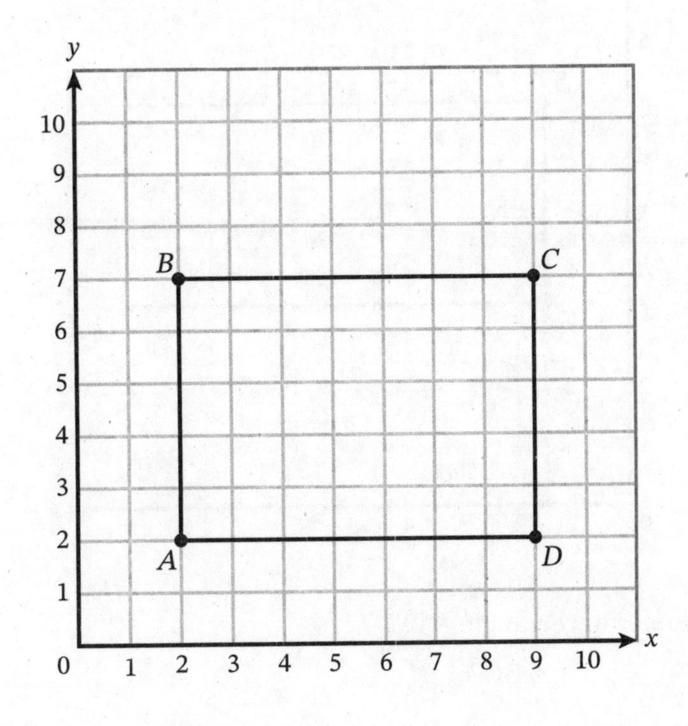

a. What is the perimeter of rectangle $ABCD$?

b. What is the area of rectangle $ABCD$?

2. Rectangle *RSTU* is graphed in the coordinate plane.

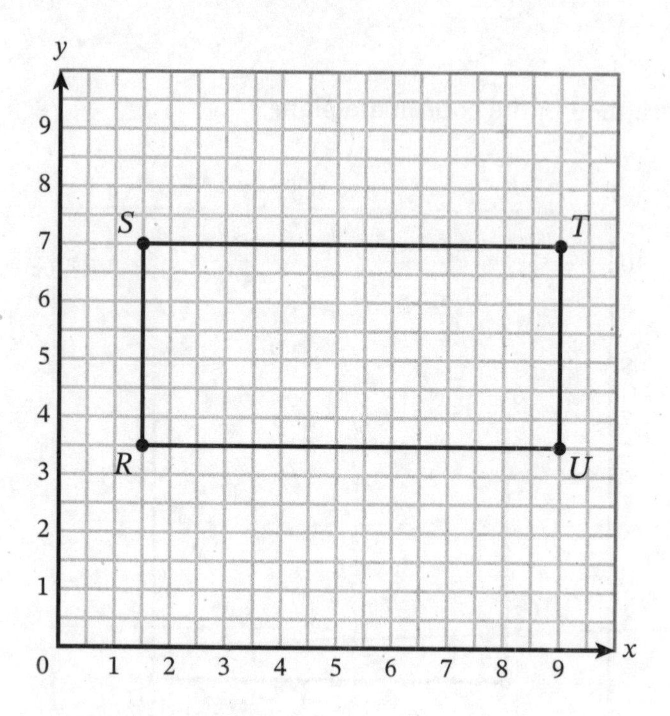

a. What is the perimeter of rectangle *RSTU*?

b. What is the area of rectangle *RSTU*?

$2 \div 3$	1
$\frac{2}{4}$	$3 \div 2$
$3 \div 3$	$1\frac{1}{3}$
$\frac{2}{3}$	$\frac{3}{2}$

2	$\frac{4}{3}$
$\frac{4}{2}$	$2 \div 4$
$4 \div 3$	$1\frac{1}{2}$
$\frac{3}{3}$	$4 \div 2$

16

Name ______________________ Date __________

1. The graph shows the number of minutes Tara practiced piano each day in one week.

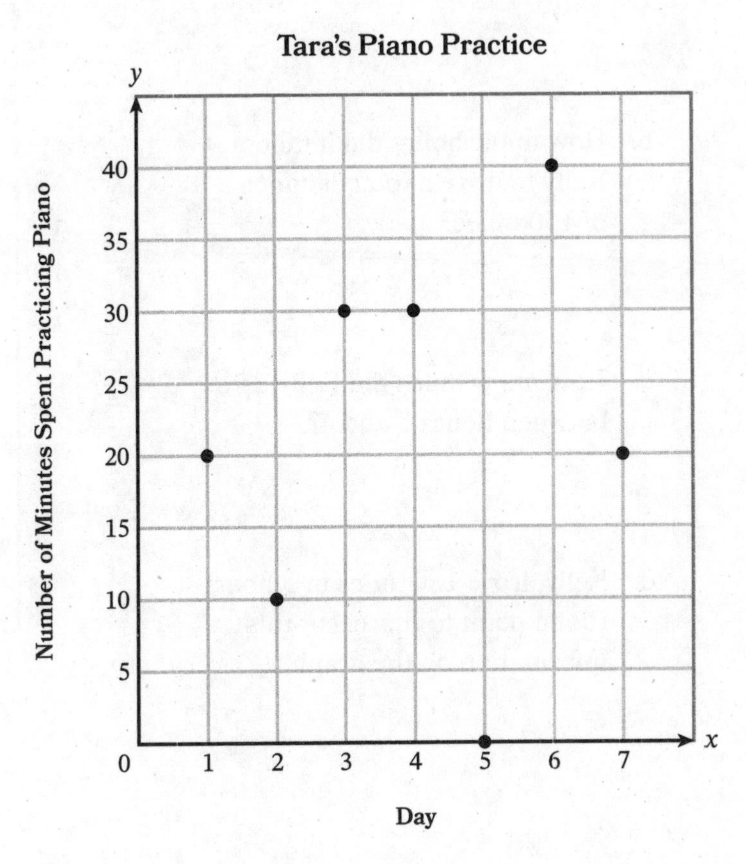

a. How many minutes did Tara practice on day 1?

b. On which day did Tara practice for 10 minutes?

c. On which day did Tara practice for the most minutes?

d. How many more minutes did Tara practice on day 6 than on day 7?

2. The graph shows the total number of miles Kelly drove after a given number of hours on a road trip.

   a. How many miles did Kelly drive in the first hour of her trip?

   b. How many hours did it take Kelly to drive a total distance of 150 miles?

   c. How many miles did Kelly drive between hours 3 and 4?

   d. Kelly drove 180 miles in 5 hours. Plot a point to represent this information on the graph.

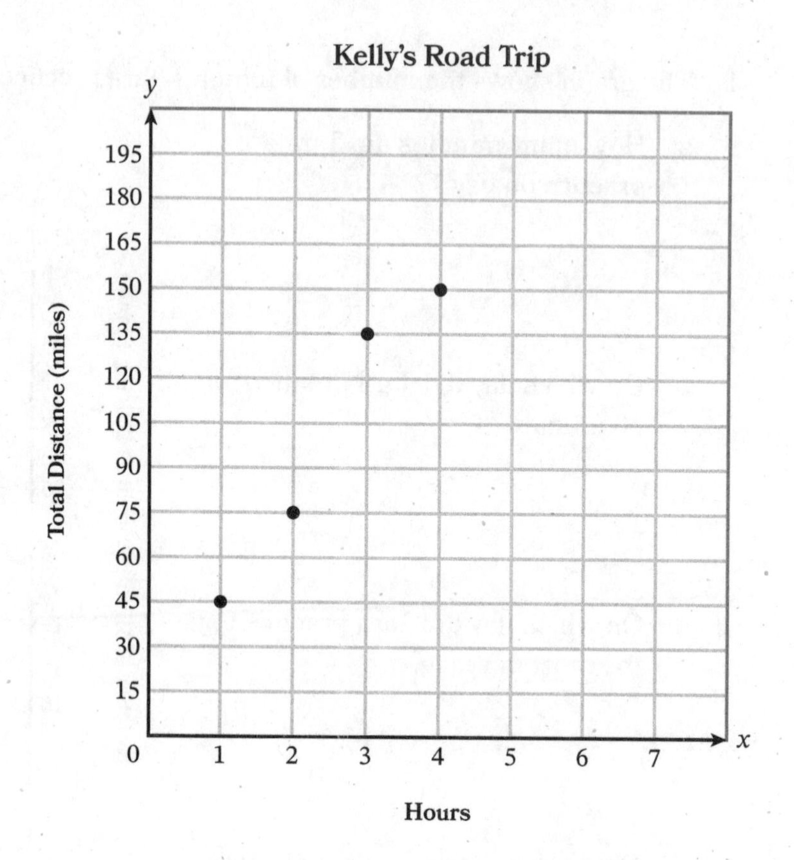

16

Name ____________________ Date __________

1. Use the graph to complete parts (a)–(f).

   a. What story does this graph tell us about how many miles Leo runs each day?

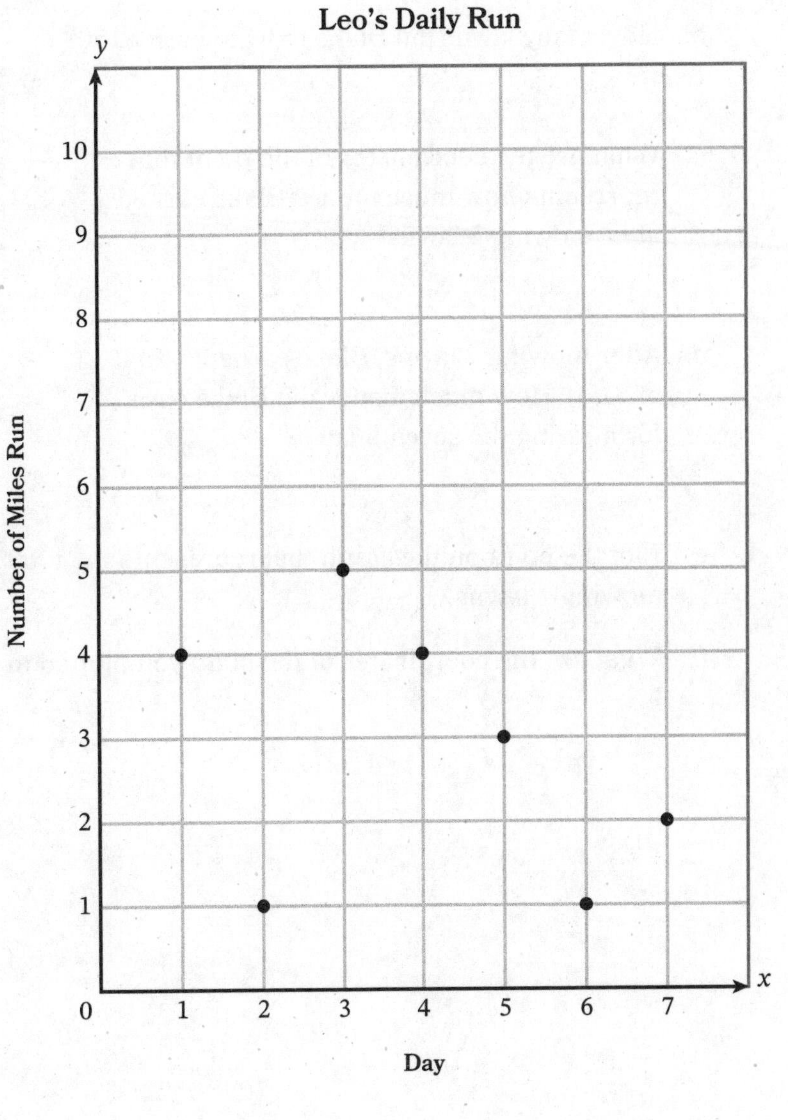

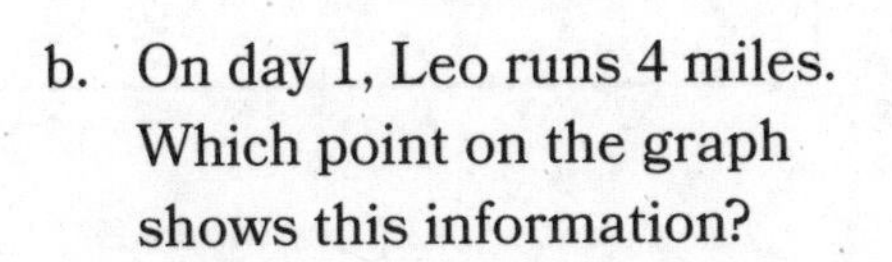

   b. On day 1, Leo runs 4 miles. Which point on the graph shows this information?

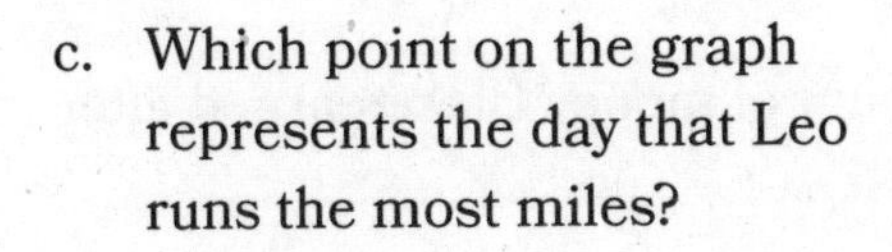

   c. Which point on the graph represents the day that Leo runs the most miles?

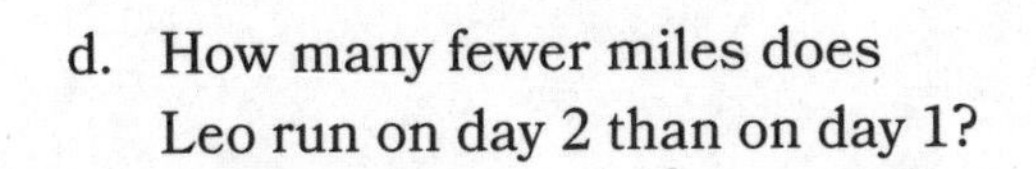

   d. How many fewer miles does Leo run on day 2 than on day 1?

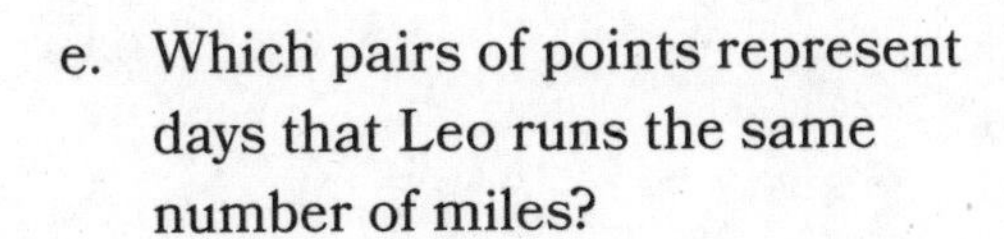

   e. Which pairs of points represent days that Leo runs the same number of miles?

      ______ and ______

      ______ and ______

   f. Riley says that Leo runs a total of 2 miles during the week because the point (7, 2) represents him running a total of 2 miles in 7 days. Is Riley correct? Explain.

2. Use the graph to complete parts (a)–(f).

   a. How much money did Blake earn for mowing the first lawn?

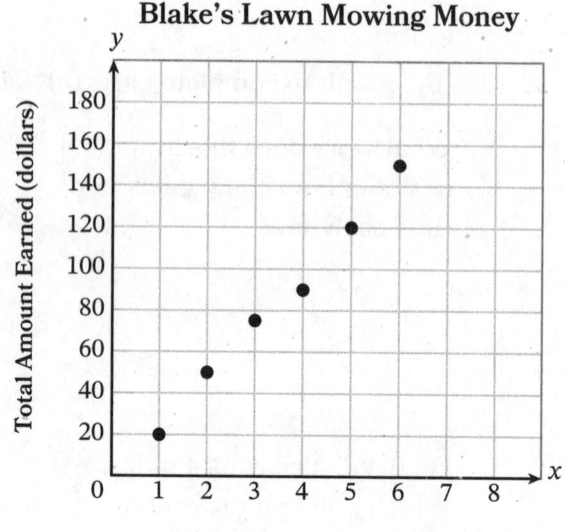

   b. How many lawns did Blake mow to earn $150?

   c. What are the coordinates of the point that represents how much money Blake earned after mowing 2 lawns?

   d. After mowing 7 lawns, Blake earned a total of $170. How much money did Blake earn for mowing the seventh lawn?

   e. Plot the point on the graph that represents the total amount of money Blake earned after mowing 7 lawns.

   f. What are the coordinates of the point you plotted in part (e)?

16

Name Date

Use the graph to complete parts (a)–(e).

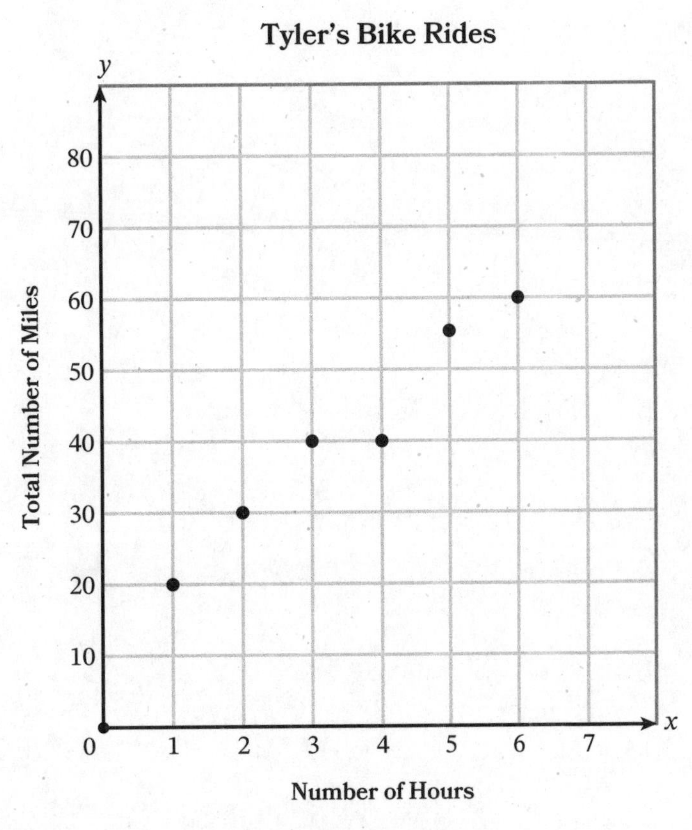

a. How long does it take Tyler to ride the first 20 miles?

b. How many miles does Tyler ride in 6 hours?

c. What does the point located at (4, 40) mean?

d. How many miles does Tyler ride between hours 2 and 3?

e. Look at the points representing the total distance Tyler rode. Why don't these points appear to lie on the same line?

$5 \div 6$	1
$\frac{6}{5}$	$6 \div 6$
$5 \div 7$	$1\frac{1}{6}$
$\frac{5}{6}$	$\frac{7}{5}$

$1\frac{2}{5}$	$\frac{6}{6}$
$\frac{7}{6}$	$7 \div 5$
$6 \div 5$	$1\frac{1}{5}$
$\frac{5}{7}$	$7 \div 6$

17

Name ______________________ Date ____________

1. Complete the table.

Word	Number of Consonants	Number of Vowels
imagine		
idea		
students		
enter		
letter		
right		
left		
data		
coordinates		
education		

2. Consider the coordinate plane.

   a. Label the x-axis Number of Consonants and the y-axis Number of Vowels. Label the title Word Data. Use the data collected in problem 1 to form ordered pairs. Plot points that represent the ordered pairs in the coordinate plane.

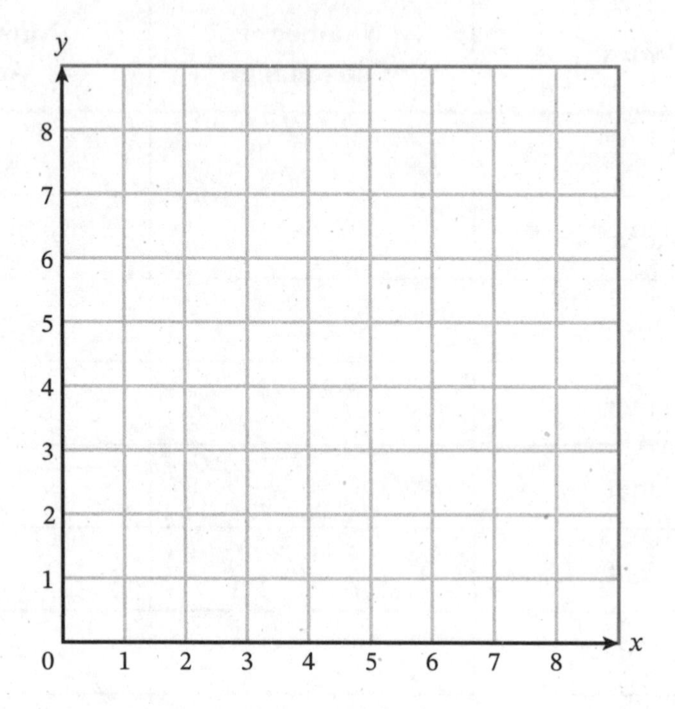

   b. Draw a dotted line from (0, 0) to (8, 8). How many plotted points lie on the dotted line? What does this tell you?

   c. How many plotted points lie below the dotted line? What do the words these points represent have in common?

   d. How many plotted points lie above the dotted line? What do the words these points represent have in common?

e. Why is the point that represents the word *idea* closer to the y-axis than to the x-axis?

f. Why is the point that represents the word *left* closer to the x-axis than to the y-axis?

g. What is an ordered pair for a point that lies on the same vertical line as the point that represents the word *letter*? What do the words these points represent have in common?

h. What is an ordered pair for a point that lies on the same horizontal line as the point that represents the word *letter*? What do the words these points represent have in common?

i. Why is the point that represents the word *education* left 2 units from the point that represents the word *coordinates*?

j. Why is the point that represents the word *students* down 3 units from the point that represents the word *coordinates*?

3. Consider the table.

   a. Write a word of each type, the number of consonants in the word, and the number of vowels in the word. Do not write the same words as the words in problem 1.

Type of Word	Word	Number of Consonants	Number of Vowels
1-letter word			
2-letter word			
3-letter word			
4-letter word			
5-letter word			
6-letter word			
7-letter word			
8-letter word			
9-letter word			
10-letter word			

b. Label the x-axis Number of Consonants and the y-axis Number of Vowels. Label the title Word Data. Use the data collected in part (a) to form ordered pairs. Plot points that represent the ordered pairs in the coordinate plane.

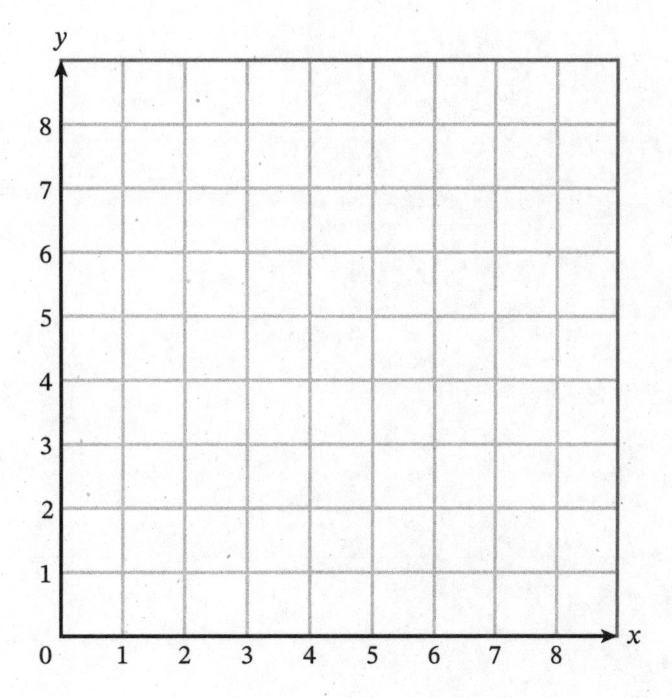

c. Based on the data, do you think it is true that the more consonants a word has, the more vowels it has? Why?

17

Name _______________ Date _______

Use the graph to answer parts (a)–(c).

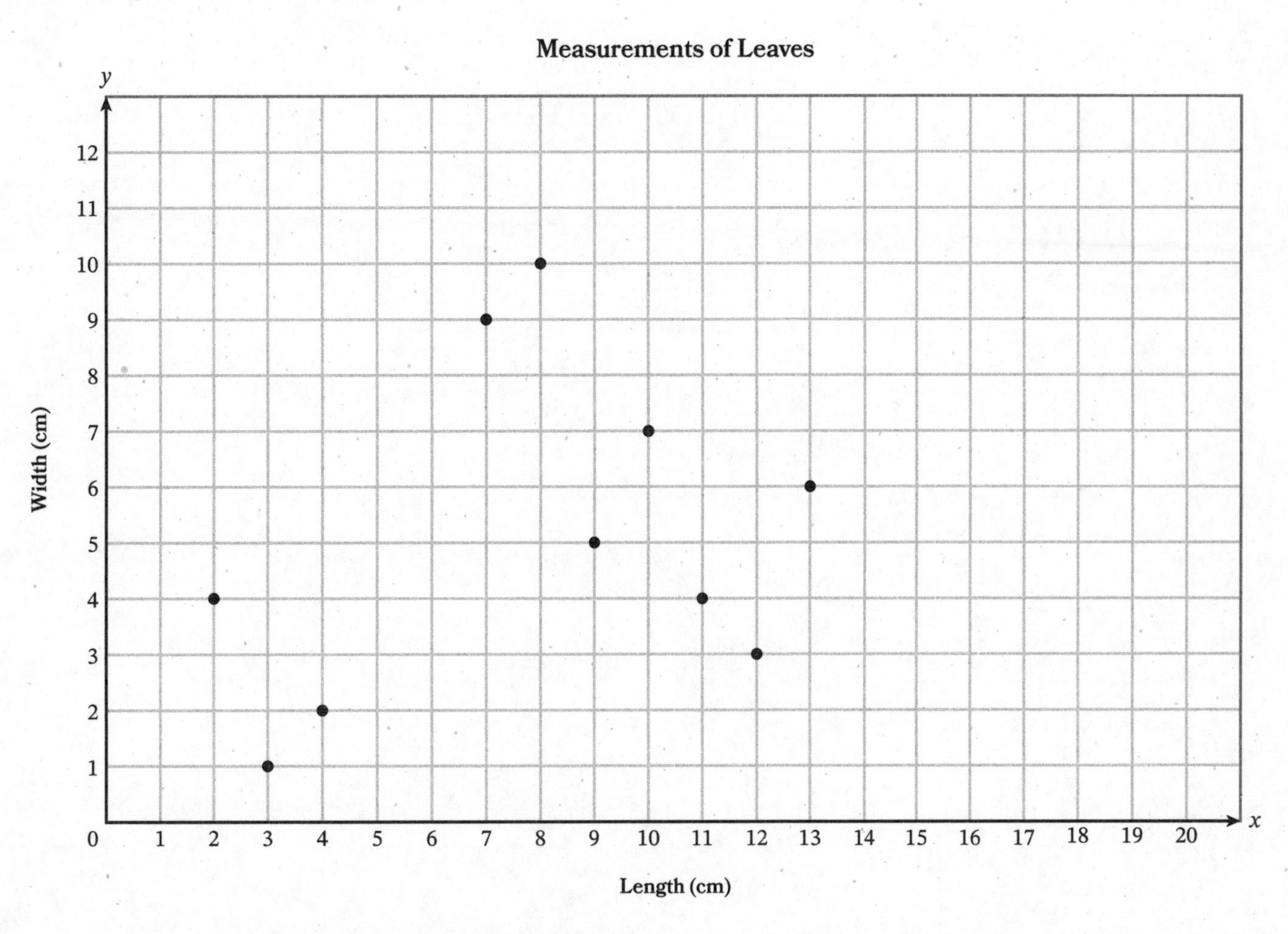

a. What does the point (11, 4) represent?

b. Are there any leaves with the same width? How can you tell from the graph?

c. Are there any leaves with the same length? How can you tell from the graph?

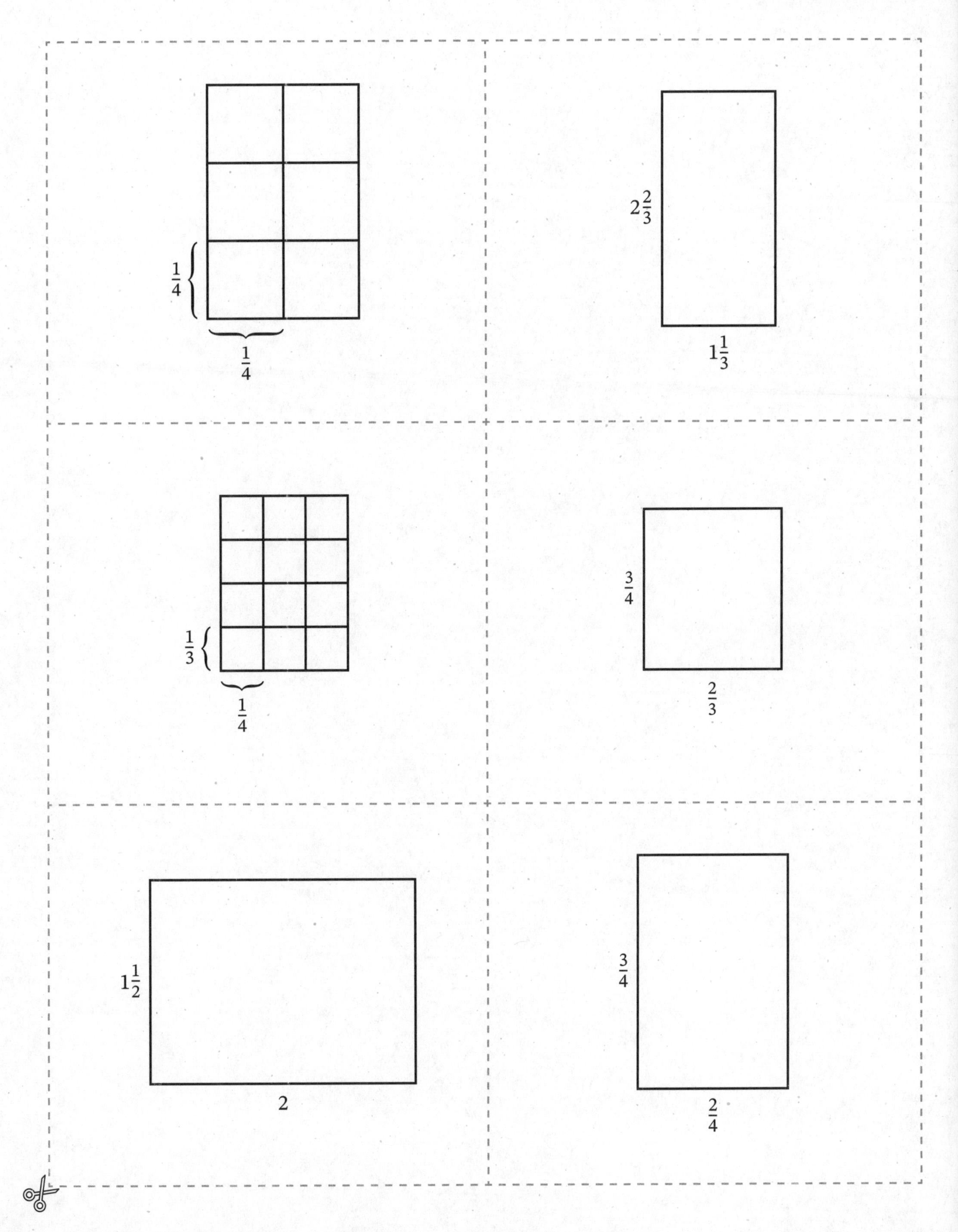
$\frac{1}{4}$
$\frac{1}{4}$
$2\frac{2}{3}$
$1\frac{1}{3}$
$\frac{1}{3}$
$\frac{1}{4}$
$\frac{3}{4}$
$\frac{2}{3}$
$1\frac{1}{2}$
2
$\frac{3}{4}$
$\frac{2}{4}$

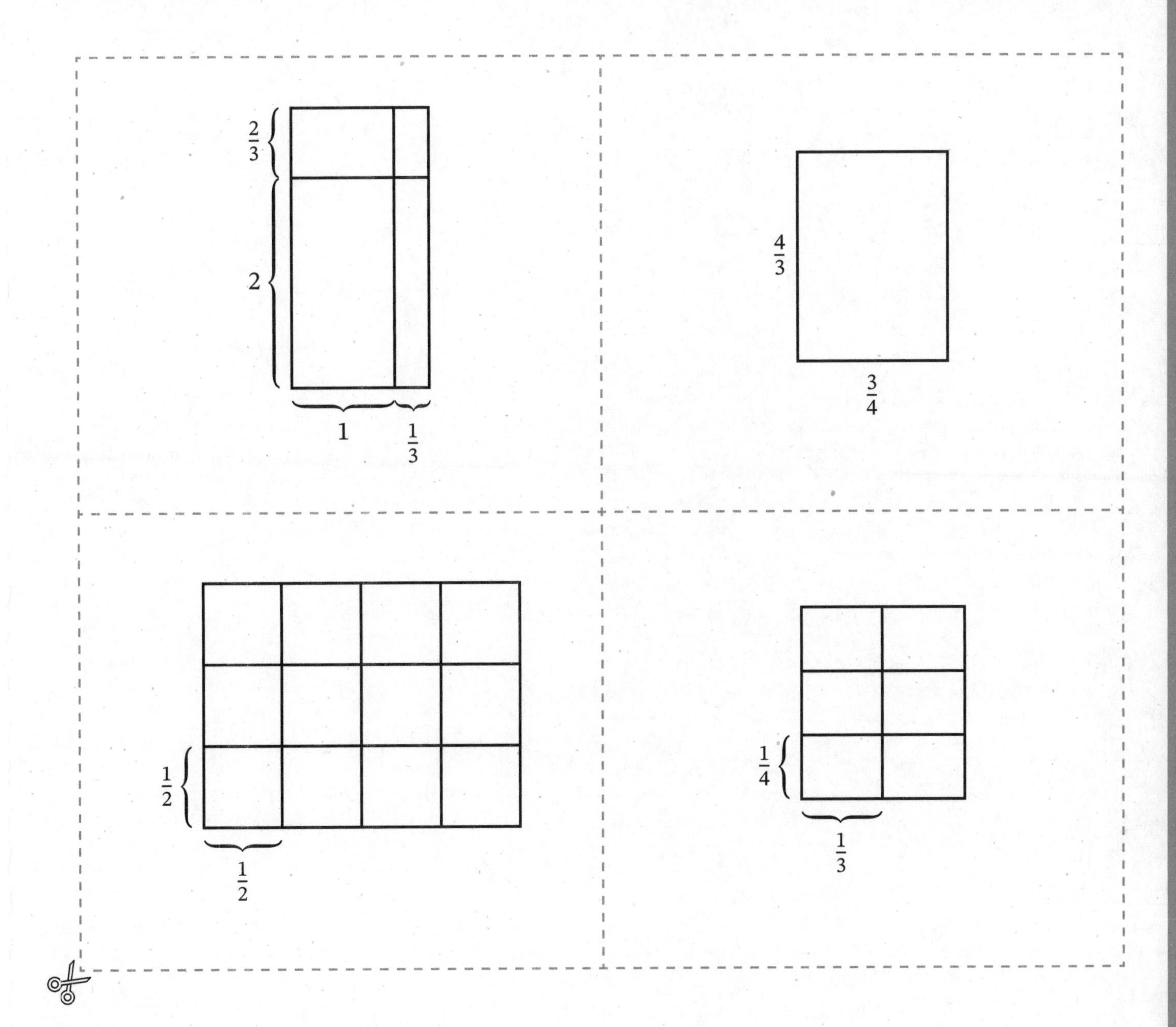
$\frac{2}{3}$
2
1
$\frac{1}{3}$
$\frac{4}{3}$
$\frac{3}{4}$
$\frac{1}{2}$
$\frac{1}{2}$
$\frac{1}{4}$
$\frac{1}{3}$

x-Coordinate	y-Coordinate	Ordered Pair

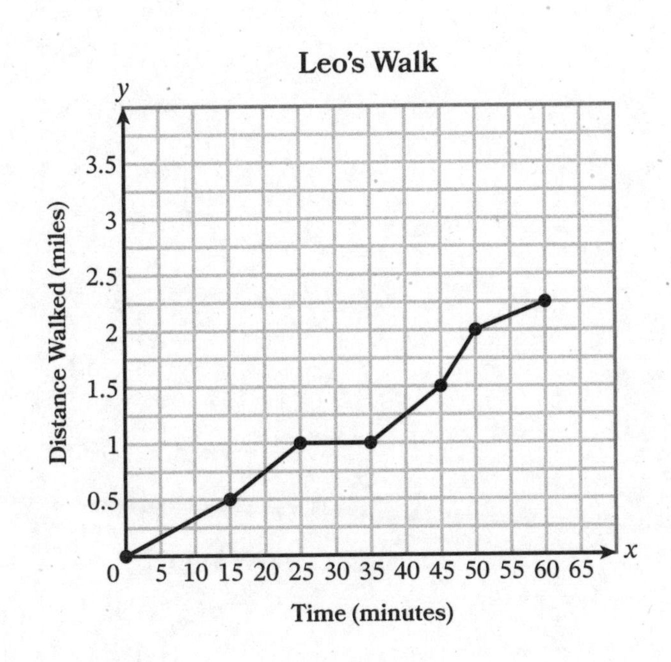

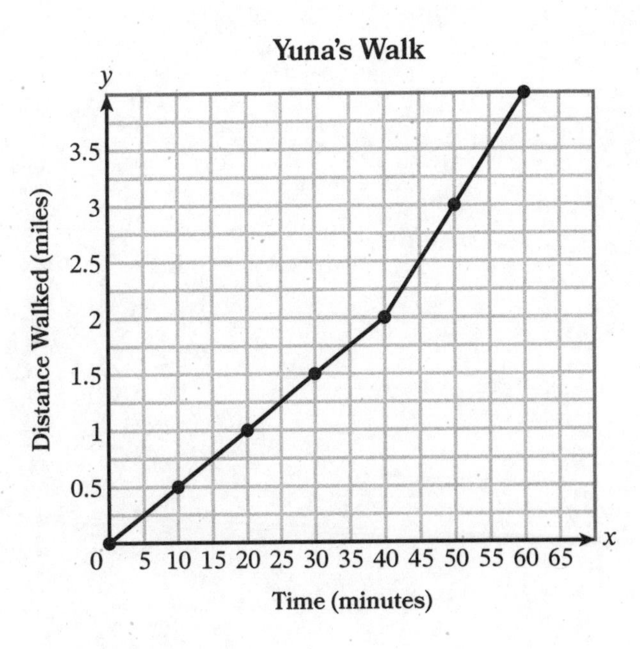

Discussion questions:

- What does the point (0, 0) represent on Yuna's graph? Does it represent the same thing on Leo's graph?
- Why do you think the line is horizontal on Leo's graph between 25 and 35 minutes?
- Did Yuna walk faster between minutes 30 and 40 or between minutes 40 and 50? How do you know?
- Did Yuna return to where she started? How do you know?
- Why don't any line segments go down from left to right?
- A line segment on Leo's graph goes through the point (20, 0.75). Do we know for certain that Leo walked 0.75 miles after 20 minutes?
- Can you use this line graph to predict how far Yuna walked after 70 minutes? Why?

Name ____________________ Date __________

1. The line graph shows the number of dogs in an animal shelter at the start of each month over a year. Use the graph to complete parts (a)–(d).

    a. At the start of which month does the animal shelter have the greatest number of dogs?

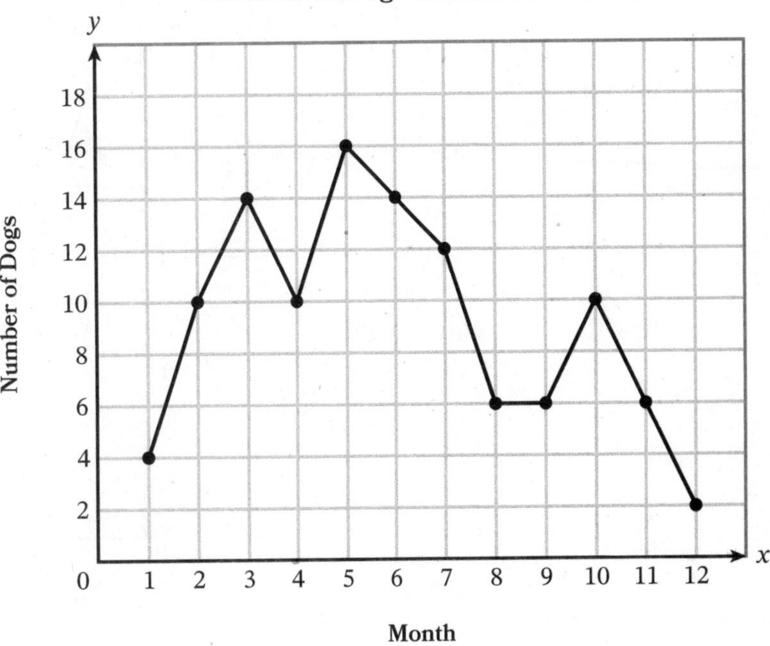

    b. At the start of which month does the animal shelter have 2 dogs?

    c. Between the starts of which 2 months does the number of dogs not change?

    d. How many more dogs are at the shelter at the start of month 5 than at the start of month 4?

2. The line graph shows the number of health points a video game character has at the start of each segment of a quest. Use the graph to complete parts (a)–(d).

   a. During which segment does the number of the character's health points decrease most rapidly?

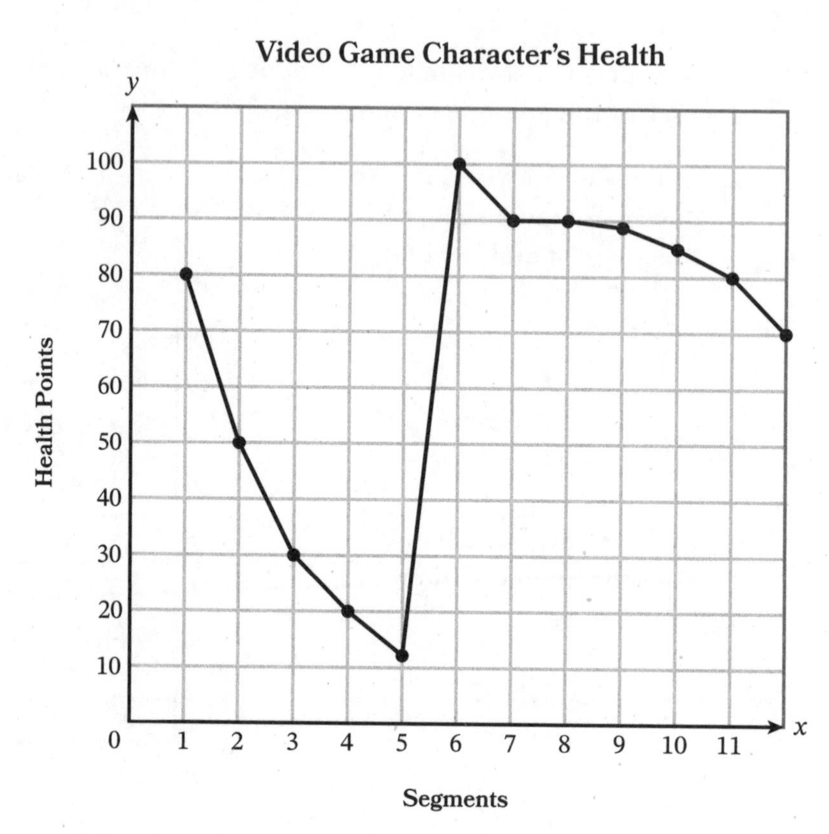

   b. During which segment does the number of the character's health points stay the same?

   c. During which segment does the character add health points? Explain.

   d. How many health points does the character use from the start of segment 1 to the start of segment 3?

3. Sasha measures the amount of rainfall during a rainstorm every half hour for 5 hours. Use the graph that shows her results to complete parts (a)–(d).

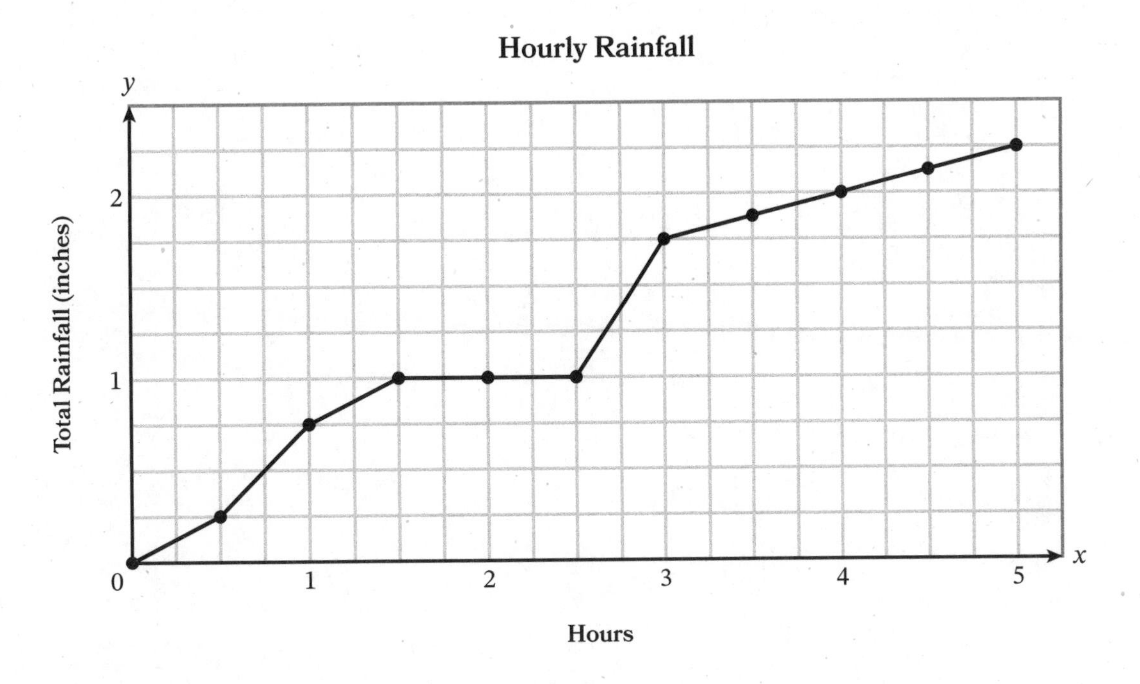

a. How many inches of rain fell during this 5-hour period?

b. During which half-hour period did 0.5 inches of rain fall?

c. During which half-hour period did rain fall the fastest?

d. Why is the line segment horizontal between 1.5 hours and 2.5 hours?

18

Name ____________________ Date ________

Use the graph to complete parts (a)–(e).

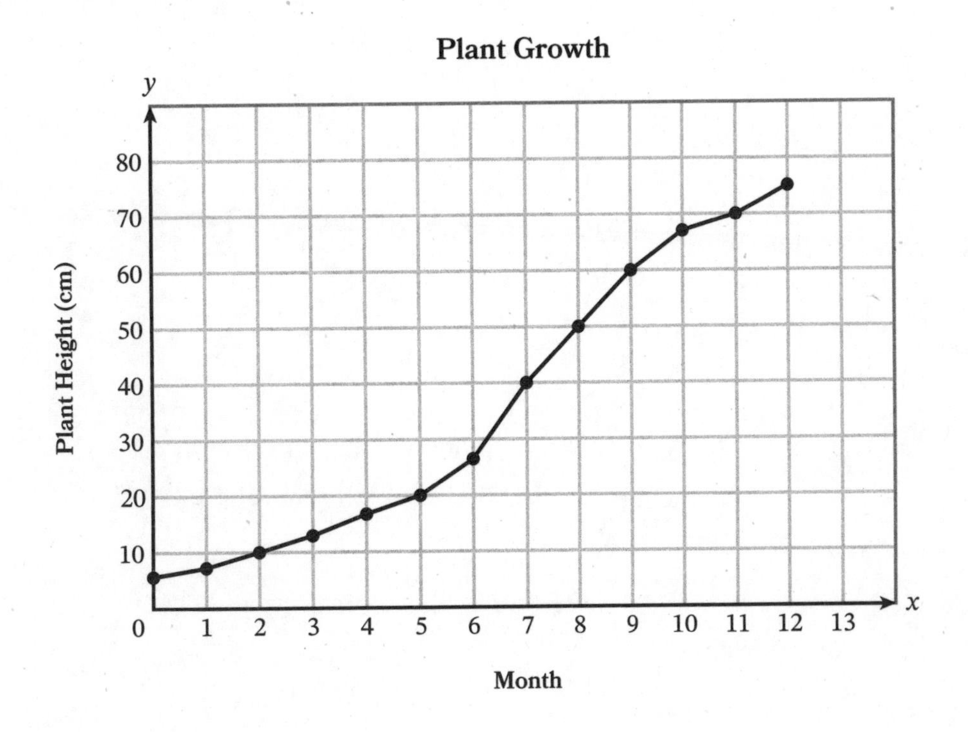

a. What story does this line graph tell?

b. Why does the graph not have any flat line segments?

c. How long did it take the plant to grow to a height of 60 centimeters?

d. Approximately how tall is the plant at month 10?

e. During which month did the plant grow the most? How do you know?

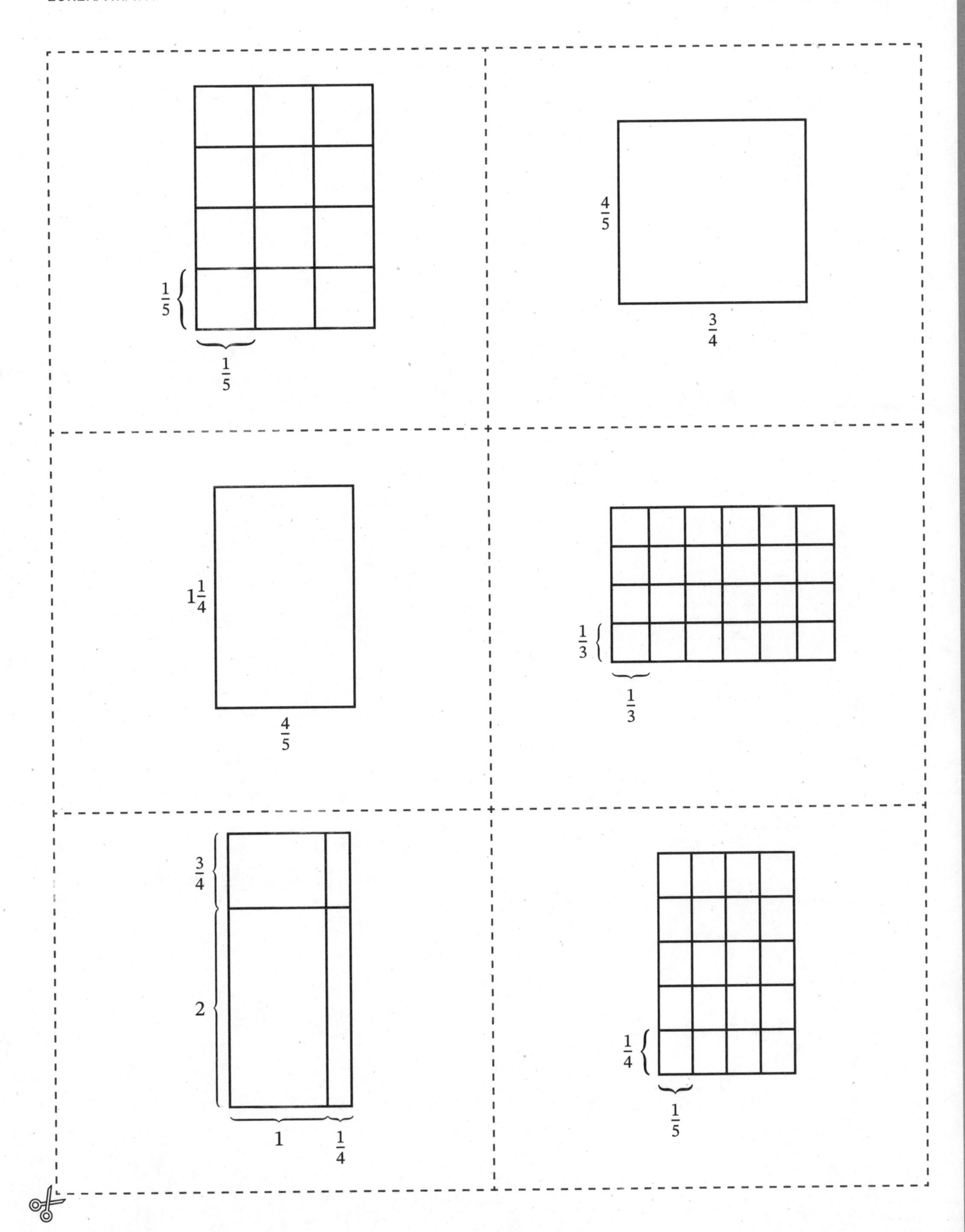
$\frac{1}{5}$
$\frac{1}{5}$
$\frac{4}{5}$
$\frac{3}{4}$
$1\frac{1}{4}$
$\frac{4}{5}$
$\frac{1}{3}$
$\frac{1}{3}$
$\frac{3}{4}$
2
1
$\frac{1}{4}$
$\frac{1}{4}$
$\frac{1}{5}$

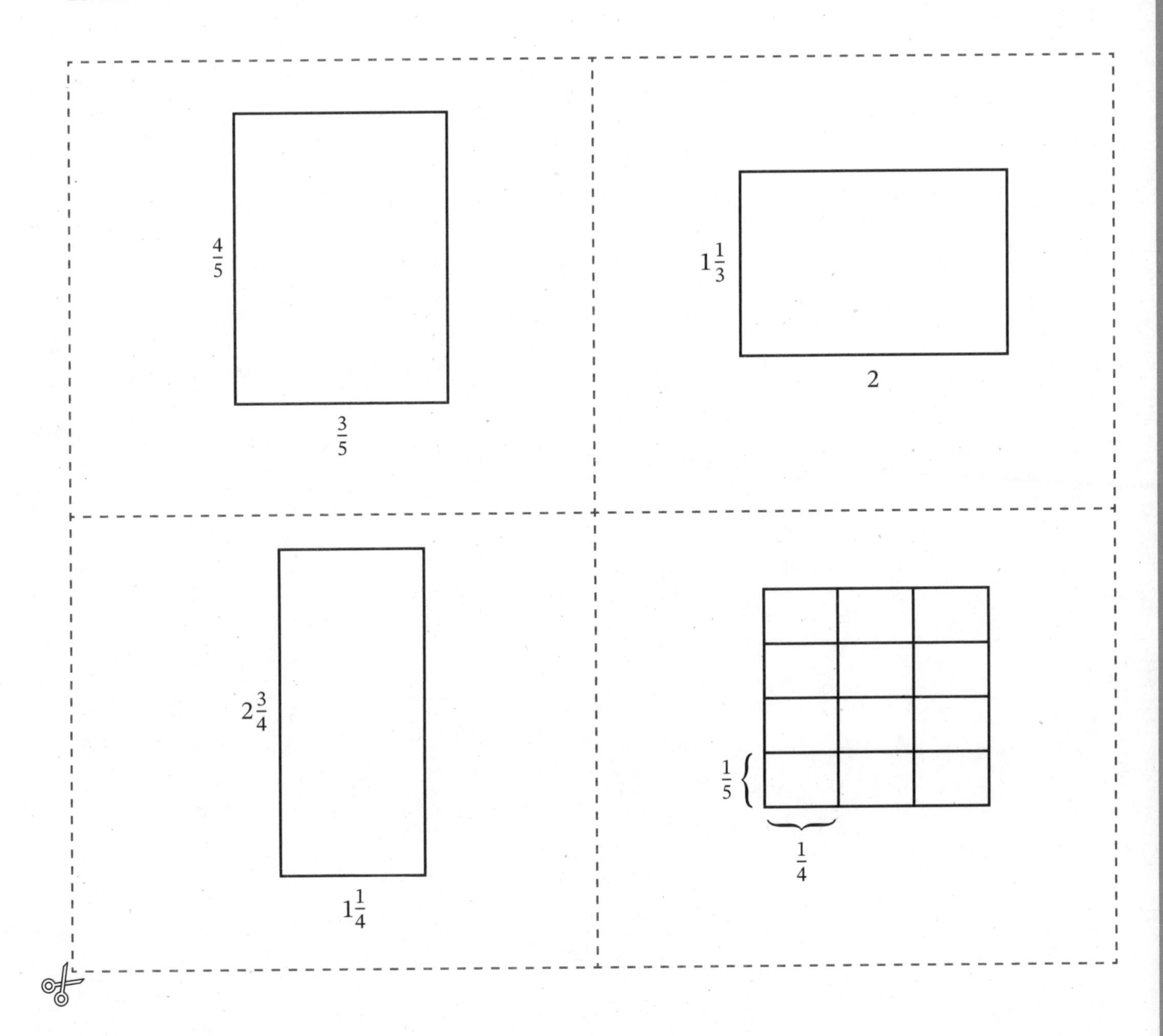
4/5
3/5
1 1/3
2
2 3/4
1 1/4
1/5
1/4

x-Coordinate	y-Coordinate	Ordered Pair

19

Name ______________________ Date __________

1. The first three iterations in a pattern of stars and hearts is shown.

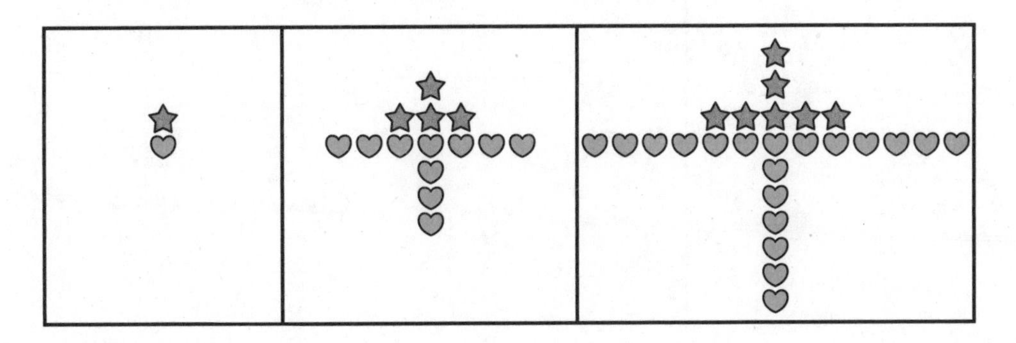

a. The table represents the number of stars and hearts in each iteration. Complete the table where x represents the number of stars and y represents the number of hearts in each step.

Number of Stars	Number of Hearts	Ordered Pair
1	1	(1, 1)

b. Use the coordinate plane to plot the ordered pairs from part (a). Label the axes.

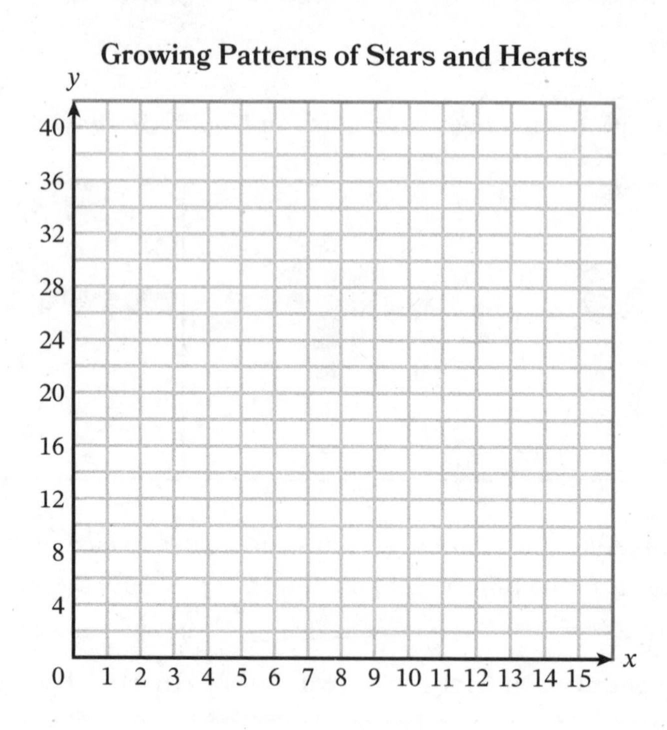

c. What is the relationship between the number of stars and the number of hearts in corresponding figures?

d. If the number of stars is 34, what is the number of hearts?

2. Square 1 is made of 1 unit square. Square 2 is made of 4 unit squares. Square 3 is made of 9 unit squares.

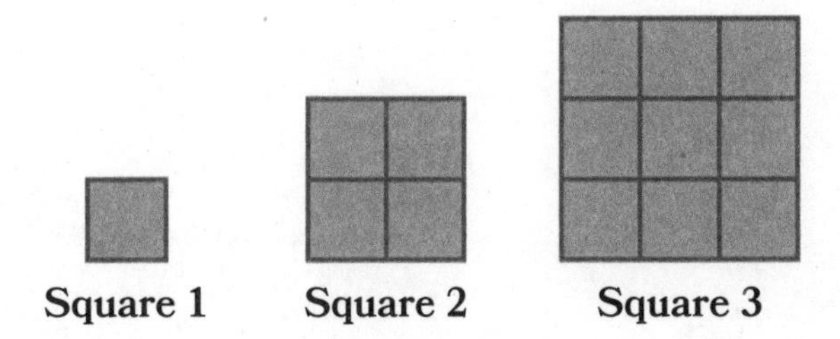

a. Draw square 4 and square 5.

b. Complete the table that shows the relationship between the area and perimeter of each square. The first row is completed for you.

Area (square units)	Perimeter (units)	Ordered Pair
1	4	(1, 4)

c. Plot the ordered pairs from part (b) in the coordinate plane.

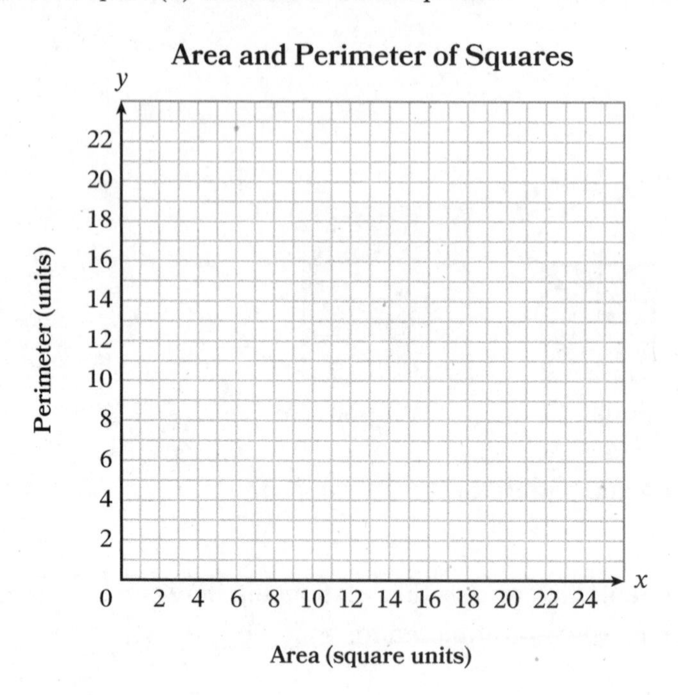

19

Name ______________________ Date ____________

The first three steps in a pattern of stars and squares are shown.

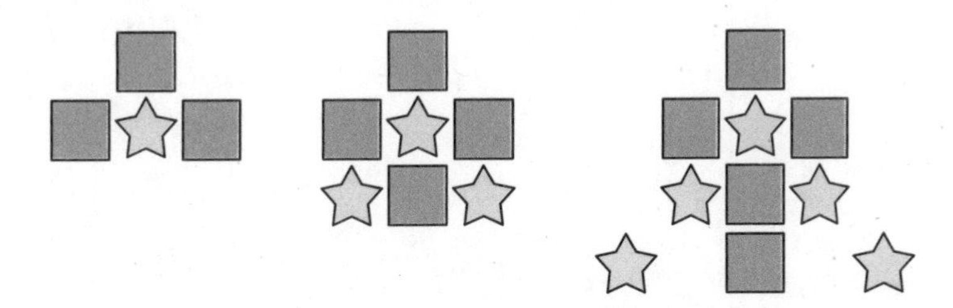

a. The table represents the number of stars and squares in each step. Complete the table where x represents the number of stars and y represents the number of squares in each step.

Number of Stars	Number of Squares	Ordered Pair
1	3	(1, 3)

b. Use the coordinate plane to plot the ordered pairs from part (a). Label the axes.

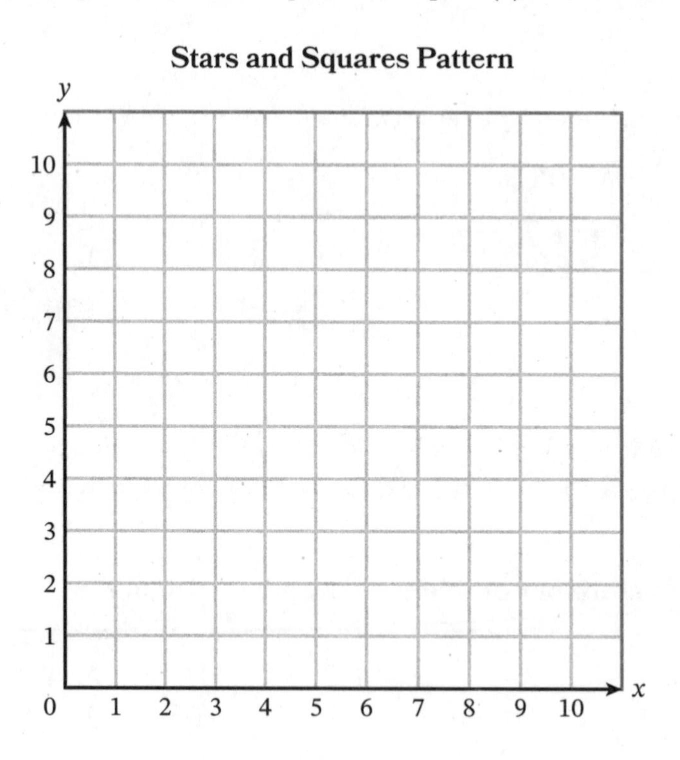

c. Describe the pattern for the number of stars.

d. Describe the pattern for the number of squares.

e. How many stars and squares would be in the fifth step of the pattern?

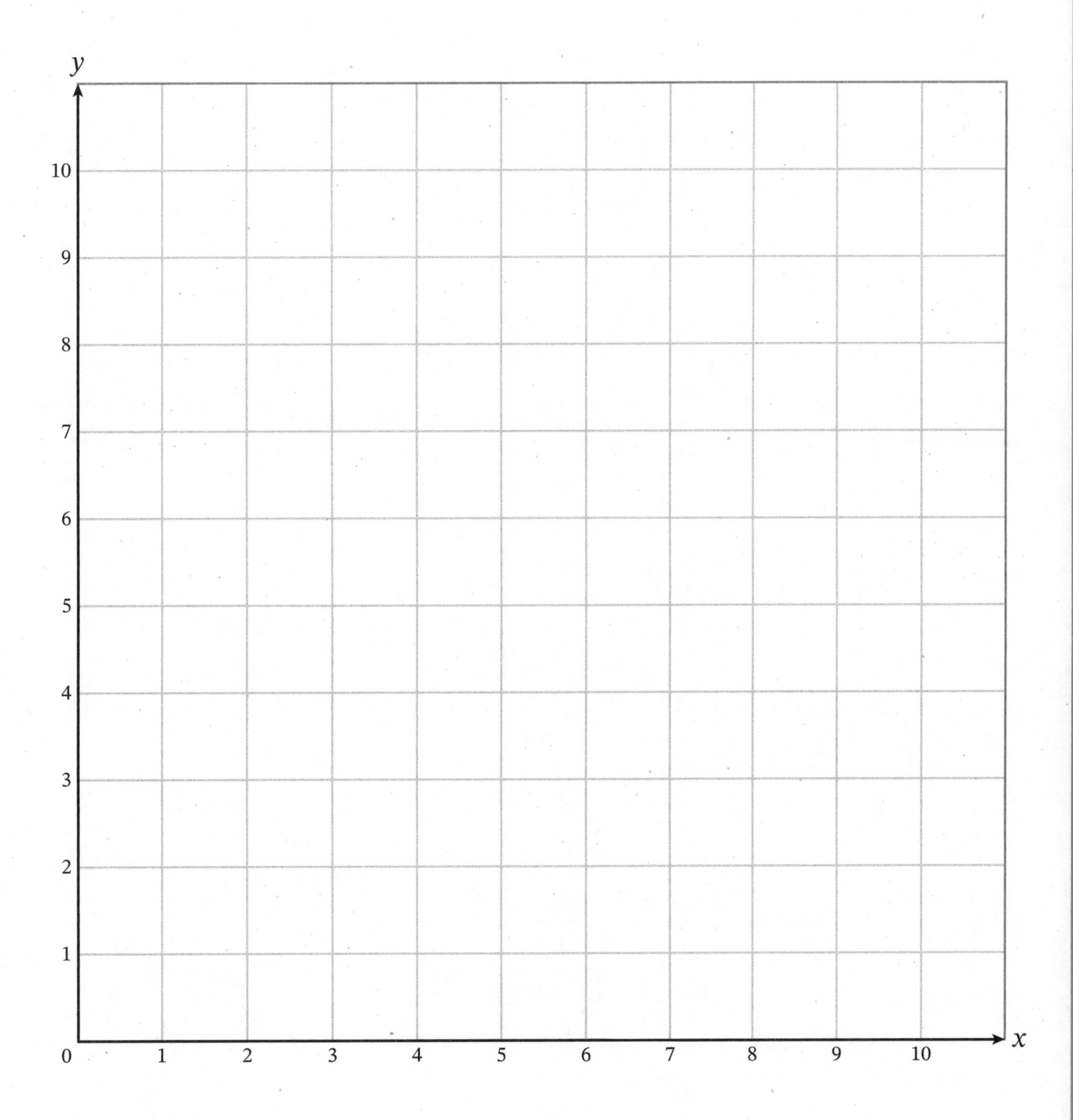
y
10
9
8
7
6
5
4
3
2
1
0
1
2
3
4
5
6
7
8
9
10
x

20

Name ______________________ Date __________

1. Write the focus question.

2. An estimate that is too low is ______. An estimate that is too high is ______. My best estimate is ______.

3. What information do you need to answer the focus question?

4. Answer the focus question. Show or explain your strategy.

20

Name

Date

Use this space to reflect on today's lesson.

Credits

Great Minds® has made every effort to obtain permission for the reprinting of all copyrighted material. If any owner of copyrighted material is not acknowledged herein, please contact Great Minds for proper acknowledgment in all future editions and reprints of this module.

Cover, Wassily Kandinsky (1866–1944), *Thirteen Rectangles*, 1930. Oil on cardboard, 70 x 60 cm. Musee des Beaux-Arts, Nantes, France. © 2020 Artists Rights Society (ARS), New York. Image credit: © RMN-Grand Palais/Art Resource, NY.; All other images are the property of Great Minds.

For a complete list of credits, visit http://eurmath.link/media-credits.

Acknowledgments

Kelly Alsup, Adam Baker, Agnes P. Bannigan, Reshma P Bell, Joseph T. Brennan, Dawn Burns, Amanda H. Carter, David Choukalas, Mary Christensen-Cooper, Cheri DeBusk, Lauren DelFavero, Jill Diniz, Mary Drayer, Karen Eckberg, Melissa Elias, Danielle A Esposito, Janice Fan, Scott Farrar, Krysta Gibbs, January Gordon, Torrie K. Guzzetta, Kimberly Hager, Karen Hall, Eddie Hampton, Andrea Hart, Stefanie Hassan, Tiffany Hill, Christine Hopkinson, Rachel Hylton, Travis Jones, Laura Khalil, Raena King, Jennifer Koepp Neeley, Emily Koesters, Liz Krisher, Leticia Lemus, Marie Libassi-Behr, Courtney Lowe, Sonia Mabry, Bobbe Maier, Ben McCarty, Maureen McNamara Jones, Pat Mohr, Bruce Myers, Marya Myers, Kati O'Neill, Darion Pack, Geoff Patterson, Victoria Peacock, Maximilian Peiler-Burrows, Brian Petras, April Picard, Marlene Pineda, DesLey V. Plaisance, Lora Podgorny, Janae Pritchett, Elizabeth Re, Meri Robie-Craven, Deborah Schluben, Michael Short, Erika Silva, Jessica Sims, Heidi Strate, Theresa Streeter, James Tanton, Cathy Terwilliger, Rafael Vélez, Jessica Vialva, Allison Witcraft, Jackie Wolford, Caroline Yang, Jill Zintsmaster

Trevor Barnes, Brianna Bemel, Lisa Buckley, Adam Cardais, Christina Cooper, Natasha Curtis, Jessica Dahl, Brandon Dawley, Delsena Draper, Sandy Engelman, Tamara Estrada, Soudea Forbes, Jen Forbus, Reba Frederics, Liz Gabbard, Diana Ghazzawi, Lisa Giddens-White, Laurie Gonsoulin, Nathan Hall, Cassie Hart, Marcela Hernandez, Rachel Hirsh, Abbi Hoerst, Libby Howard, Amy Kanjuka, Ashley Kelley, Lisa King, Sarah Kopec, Drew Krepp, Crystal Love, Maya Márquez, Siena Mazero, Cindy Medici, Ivonne Mercado, Sandra Mercado, Brian Methe, Patricia Mickelberry, Mary-Lise Nazaire, Corinne Newbegin, Max Oosterbaan, Tamara Otto, Christine Palmtag, Andy Peterson, Lizette Porras, Karen Rollhauser, Neela Roy, Gina Schenck, Amy Schoon, Aaron Shields, Leigh Sterten, Mary Sudul, Lisa Sweeney, Samuel Weyand, Dave White, Charmaine Whitman, Nicole Williams, Glenda Wisenburn-Burke, Howard Yaffe

Talking Tool

Share Your Thinking	I know . . . . I did it this way because . . . . The answer is _____ because . . . . My drawing shows . . . .
Agree or Disagree	I agree because . . . . That is true because . . . . I disagree because . . . . That is not true because . . . . Do you agree or disagree with _____? Why?
Ask for Reasoning	Why did you . . . ? Can you explain . . . ? What can we do first? How is _____ related to _____?
Say It Again	I heard you say . . . . _____ said . . . . Another way to say that is . . . . What does that mean?

Thinking Tool

When I solve a problem or work on a task, I ask myself

Before	Have I done something like this before? What strategy will I use? Do I need any tools?
During	Is my strategy working? Should I try something else? Does this make sense?
After	What worked well? What will I do differently next time?

At the end of each class, I ask myself

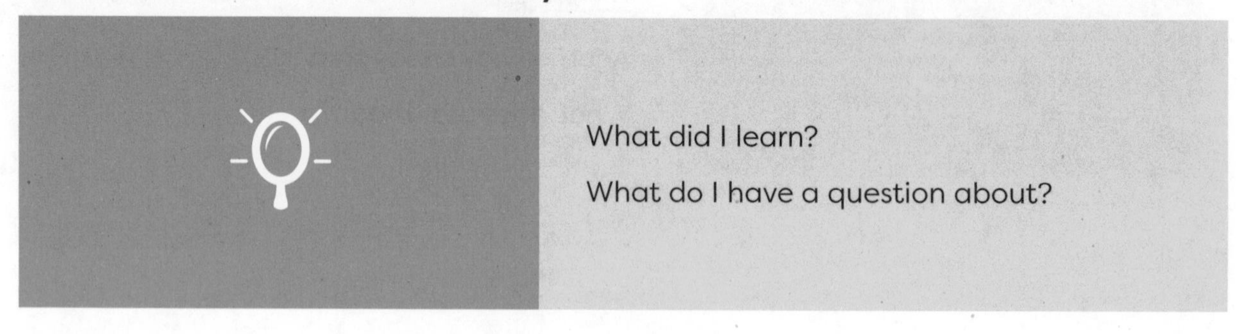